The SOLDIER of FORTUNE

The SOLDIER of FORTUNE

Exeter Books

NEW YORK

SOLDIER
of FORTUNE®

This book was compiled from text and
photographs which first appeared in *Soldier of
Fortune* magazine. The material was selected
from articles and photos which were originally
published in the magazine during its first
decade, 1975 to 1985. We are grateful for the
use of this material. We especially want to
acknowledge the textual and photographic con-
tributions of the following individuals which
were taken from the *Soldier of Fortune* magazine:
Billy Tinney, David Steele, Charles Sasser, Jeff
Cooper, Tony Bliss, Robert J. Caldwell, David
Vine, Leroy Thompson, Chuck Taylor, Fred
Reed, Don Stuber, Tim Oest, Al Venter, Ken
Pence, Sgt Gary Paul Johnston, PH2 Daryl Tuc-
ker, Lee Jurras, Bill Dempsey, Adrian Wecer,
Jeff Goldberg, Jean Dally, Jerry Ahern, Hal
Swiggett, J.D. Jones.

Compiled by: Robert L. Pigeon,
    Kenneth S. Gallagher, and Edward Wimble
Designed by: Lizbeth Nauta

Produced by Wieser & Wieser, Inc.,
118 East 25th Street, New York, NY 10010

First published in the U.S.A., 1986,
by Exeter Books.
Distributed by Bookthrift.
Exeter is a trademark of Bookthrift
Marketing, Inc.
Bookthrift is a registered trademark of
Bookthrift Marketing, Inc.
New York, New York.

Library of Congress Cataloging-in-Publication Data

Soldier of fortune.

    Selections from Soldier of fortune magazine.
    1. Soldiers of fortune.   2. Military art and
    science.   I. Soldier of fortune.
    G539.S65 1986   355'.0092'2   86-8302

ISBN 0-671-08253-1

Printed in the United States of America.

CONTENTS

Elite Men and Elite Units

Marine Recon

SOF staffer Fred Reed, a former Marine, recently went "home" to Camp Lejeune, known to Marine grunts as Camp Swampy and other even less affectionate names, to observe and report on the Second Marine Division's Second Force Reconnaissance Company. This photo series captures the rigors of RECON training.

The men of Second Force Recon, the only active-

Left: *Waiting for the mission to begin.*

duty force-recon unit in existence today, are volunteers and undergo a seven-week course at Little Creek, Va. Although based at Lejeune, they attend a wide variety of schools offered by the Army and Marines elsewhere: Ranger, Pathfinder, Sniper, Jump, HALO, Scuba and others.

The M16 is their standard weapon now but, in search of a silenced automatic firearm, they are testing the Thompson SMG, MAC-10, MP-5 and CAR-15.

Below: *Preparing for the mysteries of the deep — practicing for a hydrographic survey.*

Far left: *Recon on the beach, clothed for night ops. Ants, snakes, bees, hornets, thorns and gulleys provide the surprises.*

Left: *Swimming is sometimes the only way in.*

Below: *Marine in the bush in the jungle of Puerto Rico.*

Army National Guard Ranger students worm through a mudbath in Recondo training at Ft. Bragg. These excellent units are the only LRRP capability currently in the U.S. arsenal.

Bring Back the LRRP

Study of the U.S. Army force structure reveals a glaring gap. The combat intelligence and target acquisition capability inherent in long-range reconnaissance patrols (LRRPs) has been almost totally erased since our Vietnam experience. Despite well-researched recommendations that the gap be filled, especially in the U.S. Army in Europe, no action has been taken to provide combat commanders with the very best eyes and ears available to them—those of the LRRP. The continuing lack of an essential capability is a serious deficiency for which we will pay in blood, particularly during the earliest phases of future combat in Europe. This defect demands attention.

One of the most pressing concerns of a corps or division commander engaged in combat is knowledge of the enemy in front of him or on his flanks and how that enemy can affect his mission. We have deprived him of one of the best sources of that knowledge by not providing him with LRRPs.

On patrol, in Nam, 1967. Bill Wilkinson, Co. E, 20th Inf. (LRRP) near Kontun with ARVN 22nd Infantry Division.

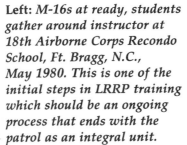

Left: *M-16s at ready, students gather around instructor at 18th Airborne Corps Recondo School, Ft. Bragg, N.C., May 1980. This is one of the initial steps in LRRP training which should be an ongoing process that ends with the patrol as an integral unit.*

U.S. Navy SEALS

SEALs are not supermen, capable of leaping tall buildings at a single bound, or tackling hundreds of enemy soldiers while chomping on a filthy cigar butt. They are human beings, with families and normal interests, who have chosen to serve their country in a unique way. They are not necessarily bigger or stronger than anyone else, nor more or less intelligent. What sets them apart from others is that they volunteered for hazardous work and had enough ambition and drive to get through programs designed to make the majority of people drop out. In many cases, their patriotism kept them going, that need to do something for their beliefs or their country. They are no different than the cop on the street or the fireman around the corner, both of whom perform a necessary, though sometimes maligned or misunderstood role in society.

Their training makes them professional, and their professionalism is in demand whenever there is an emergency. SEALs' basic training begins at the Navy's Underwater Demolition Team (UDT) 23-week school with four weeks of intense physical activity designed to condition the body through toughening runs, calisthenics, endurance or speed swims, races, competitive games, and inflatable rubber boating. The fifth week, Motivation—or Hell—week, tests recruits' mental and physical endurance to its limits

The final section of the course gives specialized training in combat and water skills. Trainees learn to use both allied and Communist weapons. They are instructed in Hwarang Do, a martial art which emphasizes the aggressive use of bare hands, knives, clubs, and other fighting weapons. Its instructors are men trained by the late Mike Echanis, Martial Arts Editor of SOF until his untimely death in Nicaragua in September 1978.

Other instruction gives SEALs mastery of boat and water pick-ups, water parachuting and land jumps, deep diving, demolition, warm water and arctic SCUBA techniques, the use of underwater propulsion craft, hand-held sonar and underwater communications equipment, hydrographic charting and reconnaissance, and first aid and survival skills.

After basic, the SEAL may be sent to the various services for additional specialized training, which ranges from HALO and language schools to commando training. Many travel to Alaska to learn underwater arctic techniques which, in the future, may become as important for SEALs' insertion in northern harbors or logistics centers as jungle survival skills were to SEALs in the Vietnam war.

Below: *Commo man catches his breath during a lull in Operation Crimson Tide in the enemy-infested Bassac River area 110 kilometers southwest of Saigon. Begun in 1962, the first SEAL cadre was drawn from existing Underwater Demolition Team forces. Eventually it was transformed into a tight elite force of hard-hitting jungle fighters for Vietnam, and other environments for contemporary missions.*

Above: *Leaping from a beached river patrol boat, a SEAL team member wears the standard garb and carries weaponry of a SEAL team in action. Note the M-203 carried by the man on the right, and the Stoner with drum magazine carried by the man next to him.*

Obituary to a Warrior: Michael D. Echanis

Soldier of Fortune Martial Arts Editor Michael D. Echanis, modern master of martial arts, who plied his skill in combat as well as in a *dojo* (gym), met an untimely death by plane crash on 8 September 1978. Echanis, whose legendary capabilities are well known to our readers, had been chief military advisor to Nicaraguan President Major Anastasio Samoza for one year, when the Aero Commander in which he was a passenger plunged into Lake Nicaragua. Also killed were Charles Sanders, Echanis' second-in-command and a close friend, a mysterious Vietnamese merc named Nguyen van "Bobby" Nguyen, and Brigadier General Jose Ivan Alegrett Perez, operations chief of the Nicaraguan National Guard.

In his role as chief military advisor, Echanis reputedly had a $5 million intelligence budget in Nicaragua. After a reputation-building year in this Central American country, Echanis, together with Sanders and Nguyen, boarded the twin-engined plane, owned and allegedly piloted by Perez. All aboard died when it fell into the waters of Lake Nicaragua. Government sources blame bad weather for the crash, but one wire service report indicated that local sources declared the weather was clear and the plane exploded in flight, then plummeted lakeward.

Was Echanis sabotaged by a jealous indigenous rival or by Marxist Sandinistas? Did his intelligence net expose too sensitive a plot. Did his unflagging honesty step on too many toes? At present it is doubtful whether we will ever know the truth about the cause of his death. In any case, Echanis was a "professional" in the true sense of the word and we at SOF mourn his death.

Mike Echanis, a former senior instructor of the U.S. Army Special Forces and U.S. Navy SEALs, was privately tutored by the Grand Master of Hwa Rang Do, Joo Bang Lee. He developed a progressive form of self-defense and survival based on combat experience and true life-and-death struggles. At the time his articles appeared in SOF, Echanis was the chief instructor for the personal bodyguards of General Anastasio Somoza, the then president of Nicaragua. It was in this role that he died. These pictures appeared in the September, 1978, issue of SOF with Contributing Editor Chuck Taylor acting as aggressor. Needless to say, Chuck is the fellow that ends up on his back.

Recce Commandos

Those military cognoscenti who have followed small war developments in the past two decades are aware of the formation of an elite fighting unit in Southern Africa: the Reconnaissance Commandos, or "Recces," as they are known locally. Though limited in numbers, these crack commando units have been compared with Britain's Special Air Forces, United States SOG elements or Rhodesia's Selous Scouts. Their work is secret and their tasks severe, often in the extreme.

While ostensibly an exclusively all-South African force, a few selected specialists from Britain and America are now attached to the Recces. But, as warned by their Commanding Officer, Commandant John More, it's not every soldier that need bother to apply:

"The Recces only accept within their ranks the very best of the best. And then they have to prove themselves in their first baptism of fire," he told SOF.

Top, left: *Volunteers receive patrol briefing from instructor. The whole program takes place in the wilds of Northern Zululand, some of the harshest terrain in Africa.*

Top, far left: *Once a day candidates are allowed to prepare a meal; in this case, a soup of biscuits and water.*

Bottom, far left: *The last leg of 20 miles with normal kit, rifle and a 70-pound sandbag. Endurance means survival.*

Bottom, middle: *Without a doubt, South Africa's Recce Commondos are one of the most elite forces in the world today.*

Bottom, left: *Recce candidates cool off during a river crossing exercise. For periods lasting as long as a month, the men are not allowed to wash and the exposed parts of their bodies are smeared with camouflage cream.*

American Merc: John Coleman

John Coleman joined the Army at age 18 in 1969 and ended up in Vietnam—by way of Germany. From his arrival in 'Nam, Coleman was not one to let grass, rice or anything else grow under his feet.

Coleman served in infantry Recon in Quang Tri and Tuy Hoa, wading paddies and pounding mountains until the end of his tour, attaining the rank of sergeant. Upon his return to The World, the Instructor's Training Course, NCO Academy and Ranger School occupied his time when he wasn't an infantry instructor at Ford Ord. Fort Lewis, Wash., Panama and Germany saw him again as sergeant of a Recon platoon until he tired of war games and decided to find a real war in a real army . . . "I became disenchanted with an army deteriorating into a bunch of officers afraid for rank and NCOs just kicking back and waiting for retirement."

Peacetime armies are hard enough for adrenaline junkies, but the peacetime citizenry is even worse. Catching a plane for Rhodesia in early 1976, Coleman was grilled by customs and a Special Branch official, before he was finally allowed near a Rhodesian recruiting office.

Coleman may have longed for a little more boredom after joining the Rhodesian Army, and a hard, disciplined, traditional military like Rhodesia's certainly disappointed more than a few Americans who swapped war stories for the benefit of the locals, flashed their medals and went over the hill to find a slightly more relaxed organization. This characterization did not make Yanks a privileged class in Rhodesia.

Not one to be easily daunted, John Coleman finished basic in the top quarter of his class. He transferred to the Rhodesian Light Infantry, completed the company-level tactics course and returned to that same tactical training course as an instructor. Coleman finished first in his Potential Officer course.

Although Coleman has little but praise for the Rhodesian Army ("I found an army that looked like an army, acted like an army, fought like an army and—best of all—had a cause worth fighting for"), he seemed to follow in the footsteps of the Rangers he admired so in Tuy Hoa and stayed in more or less constant trouble with Maj. Armstrong, second-in-command of the RLI. Considering the inevitability of change that would come with the Mugabe government and his continuing trouble with Armstrong, Coleman decided to try his luck in South Africa.

The South African Defense Forces interviewed Coleman, but they made it clear that their operation was every bit as enamored of "Salute It If It Moves" discipline as the Rhodesians, if not more so.

Jan Smuts Airport was the last piece of African real estate to be graced with the presence of John W. Coleman. In 1979 he boarded a South African Airlines jet and reluctantly headed for California.

John Coleman, merc.

Above: *Col. Charlie Beckwith, leader of the aborted Iranian hostage rescue attempt, watches ceremonies at White House honoring fifty-two freed Americans, 11/27/81.*

Right: *John K. Singlaub, Maj. Gen., U.S. Army, Ret., resigned in 1978 after criticizing Carter's decision to withdraw U.S. troops from Korea.*

How An American Formed Rhodesia's "Black Devils" Elite Armored Corps

"Black Devils" have invaded our liberated territory," screamed the Voice of Free Zimbabwe, over Radio Maputo, the capital of Marxist Mozambique.

And thus the Rhodesian Armored Corps received its unofficial official nickname from the enemy.

Why "Black Devils"? Because the commanding officer of the Armored corps, Major Darrell Winkler, had authorized his unit, which had swept through the terr base camps, to wear black jump suits to give them a unique identity.

Winkler, a field grade officer in the U.S. Army upon his discharge, who served three terms in Vietnam and three tours in Germany, refuses to give any further details regarding his military back-ground, schooling, or combat experience.

"I resigned from the U.S. Army because I was tired of it," he said. "I was embittered by our desertion of the South Vietnamese."

After his resignation from the U.S. Army, Winkler travelled all over the world and ended up in South Africa where he became interested in the Communist-sponsored terrorist war in Rhodesia. He flew to Salisbury, contacted the Rhodesian Army, and was subsequently interviewed by a board of officers. The Rhodesians took to Winkler and gave him a tour of their military installations.

Below: *Major Darrell Winkler, native of Ohio and CO of Rhodesia's Armored Corps, cuts loose with twin .30 cal. Brownings mounted on a 4×4. Nearly every vehicle in the Corps' pool has been modified to fit the specific requirements of the African bush.*

Below:
They were nicknamed "Black Devils" by a terrorist radio station.

Impressed with their discipline, organization, and professionalism as well as their officers and enlisted personnel, Winkler signed.

"The Rhodesians wanted someone to reorganize their armored unit who was highly qualified—school trained—and had a considerable amount of combat experience as well as command and staff experience with U.S. armored units in Germany and the U.S.," Winkler stated. "Furthermore, I would have an opportunity to get in a few more licks against the Communists."

Winkler was commissioned in the Rhodesian Army on 12 August 1977 and was initially assigned to a military studies course which is the equivalent to the U.S. Army's Command and General Staff College. Subsequently, he was appointed an instructor in the same school and lectured on mechanized infantry and armor operations. He returned to Colorado to arrange for his wife's passage to Rhodesia. Shortly after his return to Rhodesia, he was given the command of the Rhodesian armored corps.

Under his training, the Black Devils soon became the terrorist's most feared armored unit.

Above: *Gurkhas on maneuvers in Belize.*

Upper right: *Reverse-curved kukri has ridden at Gurkha sides for centuries. The hilt is fashioned from animal horn brought into Nepal from India. The blade is handmade by local craftsmen who use an ancient bellows and hand-powered grinding wheels. Hilt and blade are secured by a brass pommel. The scabbard is fashioned from sturdy wood and covered with tightly stretched black leather. Craftsmen of Nepal turn out 200 kukris a month for military use.*

Right: *Today the Gurkhas are a modern brigade-size force, using modern equipment and tactics.*

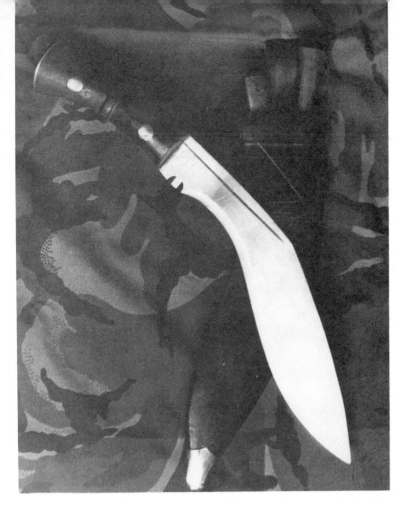

The Gurkhas

This insider's view of Britain's famed Gurkha troops was filed by David Mills, himself a former member of Her Majesty's Armed Forces, for the September, 1985 issue of SOF.

Despite an abiding popular misconception, those hard-fighting Gurkha troops of the British Army—blooded most recently in the brutal fighting for the Falklands—are not mercenaries.

They are—and have been since 1984—a respected and renowned unit of the regular British Army. These diminutive and deadly soldiers from Nepal serve in subunits of Her Majesty's Brigade of Gurkhas in assignments ranging from ceremonial guard at Buckingham Palace to security duty in Belize, a former British colony in Central America. Given their hard-earned reputation for loyalty and tenacity in combat, it's no wonder the British Ministry of Defense never hesitates to include Gurkha units in rotations to potential powderkegs.

While most Gurkha soldiers are indeed recruited from areas of Nepal that range from 5,000 to 8,000 feet above sea level (an area which encompasses just a fraction of Nepal), they can—and do—come from all parts of the country to join the ranks. The 10th Princess Mary's Own Gurkha Rifles are an example of the intense desire many Nepalese youths have to serve with the regiment. The men of the 10th are mainly Rais and Limbus recruited from eastern Nepal. Competition for acceptance is fierce. As many as 2,500 men typically apply each year for only 200 slots in the ranks. Many who fail selection for a Gurkha formation simply cross the border and join Indian Army battalions. Others, unable to face the shame of returning to their villages, drift into the slums of Terai.

Strong family connections are a feature of the 10th PMOGR as they are in other Gurkha formations. Many of the soldiers join brothers, uncles and even fathers serving in the same battalion. Such intense loyalty to a single unit is entirely understandable given the record of achievement chalked up in British military history by this celebrated Gurkha formation. The 10th Princess Mary's Own Gurkha Rifles trace their history back to 1887 when a military police battalion was raised by the government of India to protect the Kubo Valley in the western extremes of Burma. It was activated as the Kubo Valley Military Police Battalion and composed almost entirely of volunteers from Gurkha units of the Indian Army.

In 1890, a special force was formed to serve in

Burma as part of the Madras Army patrolling that turbulent part of the Empire. One of the units selected was the 10th Madras Infantry, which had been raised at Vellore in 1766. Regular officers and soldiers of the unit were replaced by Gurkhas from the Kubo Valley Police Battalion who adopted the colors and mess silver of the 10th Madras Infantry. The new unit was designated the 10th (Burma) Regiment of Madras Infantry on 1 May 1890, the date marked as the founding of the current regiment.

Although the regimental designation has undergone many changes over the years, the figure 10 has always been retained. The present title was received in 1950 from King George VI in recognition of the achievements of the regiment in World War II and to mark official affiliation with the Royal Scots (The Royal Regiment), The First of Foot.

A second Gurkha battalion was raised in 1908 and in World War I both battalions were involved in continuous combat action in Mesopotamia, Egypt and Gallipoli. During WWII two additional Gurkha battalions were raised and all four were blooded in heavy fighting in North Africa, Italy and Burma. The regiment won more awards for gallantry, suffered more casualties and spent more time in action than any other regiment of the Indian Army. After British rule in India ended in 1947, the 10th and three other Gurkha regiments were selected for continued service with the British Army. The regiment moved to Malaya in 1948 and stayed in the thick of the fighting there until the end of the emergency in 1960.

Gurkhas were heavily involved in the Indonesian confrontation of 1963 and spent 56 months in action during which they killed more enemy troops and won more awards for gallantry than any other regiment in the British Army. It was during this brutal combat that Lance Corporal Rambahadur Limbu won the Victoria Cross, Great Britain's highest decoration for gallantry in action. He is now a Captain with the 10th Princess Mary's Own Gurkha Rifles. Gurkhas also served gallantly in action against Argentine troops in the Falklands and are currently handling peacetime duties in Hong Kong, Brunei and Cyprus.

Gurkha units are mostly led by British officers who compete just as fiercely as Gurkha recruits for assignment to the Brigade. A British subaltern serving with 10th PMOGR in Belize indicated his choices for assignment after graduation from Sandhurst were listed as, (1) The Gurkhas, (2) The Gurkhas, and (3) The Gurkhas. He got his wish and the admiration of his classmates. In the chain of command between Gurkha troops and their British leaders are the officers and NCOs of the Queen's Gurkha Officers. They are responsible for purely Gurkha matters of religion, custom and other ethnic concerns.

Left: *Gurkha peering down the sights of a Bren M6. Brens are still the squad automatic of Britain's Nepalese soldiers.*

Below: *Ever vigilant, when the Gurkhas were rotated to Belize, the Guatemalan government was given clear warning of what would be in store for them.*

Roxo: The White Devil of Mozambique

The man was Daniel Francisco Roxo—an African legend. A one-man army whose head price was $100,000. A warrior whose troops of less than 100 killed more Frelimo terrorists in Mozambique than the Portuguese colonial army of 60,000. A man who strangled a leopard with his hands. A merciless killer, but a man who refused to torture the enemy. A man with an African wife and six children, who loved his home life. A man of charisma and controversy, of fact and fiction.

Roxo was born in 1933 in northeast Portugal, a tough, mountainous area, whose people are known for being hardworking, brave—and enjoying good wine. Roxo emigrated to Mozambique in the early '50s, and worked there first as a civil servant. He helped construct a railway line, and finally became a big game hunter in Niassa Province, northwest Mozambique. Big game hunters are a special breed, perhaps an endangered species in Africa. Roxo had to change his target when Frelimo (a communist-supplied Mozambique liberation front) started terrorizing the countryside in the early 1960s. His deadly aim focused on those who destroyed his live-

Below: *Roxo wearing his first and second (wreathed) Cross of War, Portugal's highest award for bravery.*

lihood, Africa's game. He grew to hate those he regarded as the cause of it all—Frelimo.

We Die As We Live

Roxo offered his services to the Portuguese army, first as tracker, then as leader of an irregular band of 90. He was the only white man in the unit. The "white devil of Niassa" demanded discipline worthy of the highest military traditions. He combined modern military methods with guerrilla tactics. His followers were desperadoes, old trackers, fellow hunters, even men recruited from Frelimo ranks. They went untracked because they wore captured enemy boots. They knew the operating procedures of the enemy and gave no quarter. They were so familiar with enemy weapons that they could immediately take them up and use them.

Roxo defined his war and his territory. He always knew the exact strength and location of the enemy before tracking. He refused to operate outside his own province, where he knew every inch of the terrain. He was ruthless and only took prisoners for interrogation. Discipline was immediate, but Roxo always disclaimed the story that he summarily punished his men or even executed them. Disobeying orders certainly meant more than a dressing-down or a tongue lashing. Troops had to proceed not less than 10 yards apart to minimize losses in case of an ambush. If the rules were broken, Roxo warned only once.

He said, "It's normal to be brave. To risk danger is itself a sign of bravery. We all choose the manner in which we live. It is often the way in which we will die." Prophetic words. . .

In the fall of 1976, he stepped on a landmine and died.

Below: *Roxo in a maize field giving instructions to his men.*

Top: *Teams were a mixed force, usually three Americans and six to nine indigenous troops.*

Bottom: *A naval operation at DaNang, an air-wing detachment at Nha Trang and a training element at a place called Bearcat, fourteen miles east of Saigon, were established.*

MACVSOG was established in 1964 to train, advise and logistically support the South Vietnamese "special Branch" LLDB — later known as the Special Exploitation Service (SES) — in the conduct of covert missions in Southeast Asia. Among the missions carried out by MACVSOG were the monitoring and interdiction of communist logistics on the Ho Chi Minh Trail in Laos, PT-boat raids on North Vietnamese coastal installations, an ongoing attempt to shut down the North Vietnamese fishing industry, escape and evasion nets for downed airmen, raiding and ambushing trains, convoys and coastal-defense units in the North.

MACVSOG

Top Secret: Burn after reading or ingest.

Much of what you are about to read is covered by the National Security Act. Documents which give details on operations and projects described herein can be traced by looking up the index numbers listed in the Military Assistance Command Vietnam (MACV) history in the Pentagon.

However, the MACV history is a classified document.

Although most Vietnam veterans, war correspondents and students of the conflict are familiar with the acronym SOG—via rumor-control central, fleeting contact in 'Nam or short references that have appeared over the years in newspapers, magazines, books and movies—allusions to the group's activities have been few.

The acronym SOG stood for Studies and Observation Group; the acronym itself was condensed from the full-blown MACVSOG: Military Assistance Command Vietnam Studies and Observation Group. Unofficially it was also known as the Special Operations Group.

SOG was organized in February 1964 by Gen. Paul D. Harkins, Commander MACV. Its basic mission was to conduct covert operations in the de-nied areas of Vietnam, Laos and Cambodia. Activities within the sphere of SOG's mission included: guerrilla warfare, subversion, sabotage, escape and evasion, direct-action missions, black and gray psychological operations and some operations best described just as unconventional warfare.

For a number of obvious and valid reasons, SOG from the beginning depended heavily on the U.S. Army for its personnel. But since it also drew troops from the Marine Corps (Force Recon), U.S. Navy (SEALs) and Air Force (air crews), and drew headquarters staff from all four services, it was by definition a Joint Unconventional Warfare Task Force. Since the Vietnamese were also involved, it could be called a Combined Unconventional Warfare Task Force. In addition, SOG used American civilians from the Central Intelligence Agency (CIA) and United States Information Agency (USIA), as well as civilians from other countries.

Depending on whose estimate one accepts, SOG during its heyday had 2,000 to 2,500 Americans, 7,000 to 8,000 Vietnamese (North and South), Nungs, Cambodians, Laotians, Chinese and others involved in its myriad operations or projects.

As one former SOG troopie put it: "An octopus would wish that it had so many tentacles."

Salvadoran Naval Commandos

Four years ago, the Salvadoran Navy decided a Naval Commando force would be better suited to fighting an insurgency than conventional Marine formations. (The Navy now believes a Marine battalion is essential, too.) In August of 1982, the Naval Commando unit was formed with 60 men just back from infantry training at Fort Benning or Panama. Today the Naval Commandos number 330 men, including 12 frogmen, 90 base security troops, and 110 men who regularly man the weapons aboard Piranhas and other high-speed patrol boats. Aspiring to match the daring of U.S. SEALs at Grenada or the finesse and lightning-striking power of Marine Force Recon in Vietnam, El Salvador's Naval Commandos are determined to keep communist guerrillas from establishing beachheads on the coastal areas of their homeland. The Naval Commandos regularly prowl mangroves, coconut forests and beaches in eight- to 15-man teams, ambushing guerrilla columns and raiding rebel encampments.

Although the occurrence of close combat has been low, the Naval Commandos have badly hurt the guerrillas. Just the threat of Naval Commando First Company, put it all in perspective. "We use guerrilla tactics. Our action is mostly psychological. The guerrillas never know where we are, but they know we're there. Even when we're not there, they are thinking about us. The guerrillas are scared."

Top left: *Displaying M-16s taken from "Gs". Many captured weapons have serial numbers indicating they were left behind by the U.S. after the Vietnam war.*

Bottom left: *The Naval Commandos regularly prowl mangroves, coconut forests and beaches in eight- to fifteen-man teams.*

Below: *In August of 1982, the Salvadoran Navy formed the Naval Commando unit aspiring to match the daring of the U.S. SEALS and Marine Force Recon. There are now 330 men, including 12 frogmen, 90 base security troops and 110 men who man the weapons aboard Piranhas and other high-speed patrol boats.*

Right: *The Swiss Army is basically organized into three field corps, each consisting of a mech division, a field division, a grenz (border) division and a mountain corps. There is also one formation that ranks with the elite of the world: shock troops of the Grenadier Paras. This photo shows their unique cammo smock and helmet.*

Below: *Swiss males between the ages of 20 and 50 undergo periodic training and are part of the military reserve.*

Right: *Infantry grenadier's collar insignia.*

Right: *Grenadier parachutist's wings are cut out and worn on the left breast.*

Swiss Shock Troops

Despite a population of only about 6½ million, the Swiss Confederation can field one of the most effective military forces in the world. Like Israel, Switzerland relies on an extremely effective reserve system to field an army 600,000 strong in a national emergency. Of this number, however, only about 2,500 are professional soldiers. The 40,000-strong Air Force contains another 3,000 professionals.

Although the quality of all Swiss troops is very high, there are a few specially trained "elite" troops who are virtually unknown outside of Switzerland. Within their infantry regiments, for example, are grenadiers trained in demolitions, flame throwers and other special skills.

The Swiss special forces unit—in the sense of the SAS or U.S. Special Forces—is the Company of Parachute Grenadiers assigned to the Swiss Air Force. The first Swiss school for Grenadier Parachutists was run in 1970. While building the company up to full strength, schools were run in each of the next four years, but between 1974 and 1982 schools were run only every other year. Since 1982 schools have been run yearly again. In addition to static line and free-fall parachute techniques, Ranger/LRRP training is also part of the course. Pistol marksmanship is also stressed since the Grenadier Parachutist often jumps with just a pistol. Demolitions, survival, intelligence gathering and other skills normally associated with special ops are also taught.

Each year about 250 Swiss apply for grenadier para training, of which only the top 24 are accepted for the recruit school. Of these 24, about eight usually successfully complete the training and receive their wings. This rigorous selection and training program makes the Grenadier Parachutists one of the world's truly elite units. These newly qualified paratroopers replace those members of the company who, having turned 33, must revert to the *Landwehr*. Officers and NCOs can remain a few years longer, but only the company commander may be over 40.

The Grenadier Parachutists are normally identifiable by their fiber jump-helmets and their jump boots. They wear standard Swiss camouflage. The only insignia normally worn is the company shoulder-slide bearing the number 17. It is worn on all uniforms by NCOs and below. Parachutist's wings and collar insignia are worn on mess and service dress.

The number of Grenadier Parachutists remains small, and even their existence has remained unknown to most. Any invader of Switzerland, however, will soon know they exist; the death and destruction in his rear will make him well aware of the Grenadier Parachutists.

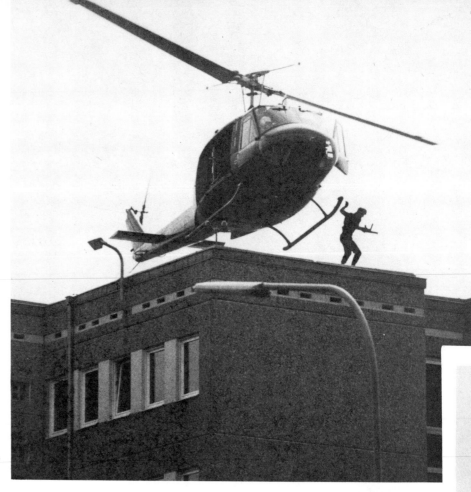

Right: Helicopter insertion is a major part of GSG 9 training.

GSG9: Anti-Terrorist Elite

Queer as it may seem, there are those who actually try to justify terrorism. They claim that it is the weapon of the weak against the strong, and maintain that it is a legitimate act of war. This is ideological solecism, and to give it a moment's credence is a sin in itself. If one has a grievance his target is its author. To strike at a third party in order to coerce your adversary is worse than criminal—it is foul. The terrorist thus places himself outside of any considerations of respect or pity. He is worthy of all the mercy we afford the rabies virus.

In general, we wring our hands and whimper. But not, however, *all* of us. Good men, worldwide, are growing angry. And with the anger may come the will. The hard line is indeed unfashionable at this time—at least in the Western World. But the case is still not hopeless. Certain groups are now organized to take suitable action against this hideous social chancre.

Pre-eminent among these is GSG 9 *(Grenzschutzgruppe 9)*, the special commando of the German border patrol, conceived at the Munich Olympic Games disaster of 1972, and consecrated at Mogadishu in 1977. This unit is now fully prepared and fully tested in action—a bright, sharp, and finely tempered sword ready for use wherever the Federal Republic of Germany may send it.

Below: *Armed and ready, GSG 9 team awaits assignment.*

A high sidekick and timely block prove mightier than a "butt stroke" from a work-out partner.

ROK Special Forces: The Tigers of Korea

Their berets are black, and their intentions toward the communists who lurk across the uneasy border are the same hue.

Occasionally, you see these men on the streets of Seoul, but only on the weekends and comparatively few at any one time.

These men are easily the most distinctive troops in all of Korea's military forces, dressed in their camouflaged jump suits with the para wings worn proudly on the left breast. These are the ROK Army Special Forces, and when "the balloon goes up," they will be the first to carry the message of miscalculation to the doorstep of Kim Il-sung, the commie king of North Korea.

It is not easy for the average American to realize the import of the tactical situation that faces Korea 24 hours a day, year in and year out. As the years roll past since the cease-fire was called 24 years ago, the situation has intensified. The North Koreans have steadily added to the muscle of their massive war machine, which has put that country some two billion dollars in debt to all of the nations rash enough to trade with them. 550 tanks, nearly a thousand planes, and in the Kaesong area (just over the DMZ) alone, 23 Y-shaped long-range gun bunkers, each 80 meters long, with 42-inch thick reinforced concrete walls and twenty feet of earth fill, point at the heart of Seoul like an assassin's dagger, leaving no doubt of comrade Kim's long-range intent.

ROK on parade.

The DMZ border that runs across the Korean peninsula from East to West is guarded by thousands of troops, barbed wire, mine fields and radar, and is fairly secure, although two infiltration tunnels have been intercepted 120 feet under the solid granite of the border, and the presence of 10 more tunnels is known and duly marked on the "I" Corps commander's battle map.

The really great problem where communist infiltration is concerned, is the long and convoluted coastline of Korea where agents of the North can be dropped in by boat or sub, making an "end run" around the heavily-guarded DMZ.

This is where the mission of the Special Forces gains importance. Teams of tiger suited troopers are trained to pursue the intruders with the instincts of hunting cats, and God help their prey when contact is made!

Continued provocation has become a way of life for the tough-minded Koreans, and as long as the ROK Special Forces keep themselves honed to their usual razor sharpness, snuffing out terrorist incursion will be just another exercise.

Right: *"Tallyon Chu" forging posts give the troopers' fists a good work out.*

Robert K. Brown: Professional Adventurer Extraordinaire

Bob was born in Monroe, Michigan on November 2, 1932 but grew up mainly in Indiana. His father was a steel worker and his mother taught school. In 1950 he returned to Michigan and attended for two years the State University there before transferring to the University of Colorado in Boulder. In this small college town his participation in collegiate boxing and rodeo-riding made him a well-known figure. He also began a long succession of odd jobs that cast him as a cowboy, armored-car guard, trail-crew foreman, fire fighter, hard-rock miner, logger, carpenter photo-journalist and, as we know him now, publisher.

After earning a history degree he enlisted in the Army for a hitch that lasted from October 1954 until September 1957. As an officer in the intelligence corps he gained vital experience for the work his destiny had in store for him in later years.

When this first hitch in the Army ended he returned to Boulder to obtain his master's degree in political science. The cold war was heating up and Bob became fascinated with low-intensity guerrilla conflicts that were surrogates for the larger ideological struggle between east and west. He backed the bearded Revolutionary Fidel Castro in his budding effort to free the people of Cuba from the tyrannical, right-wing regime of Batista.

For the next several years Brown's flirtation with the sidelines of this revolution led him to a brief career as a gun-runner and a trip to Havana for a forlorn attempt to meet Castro.

He returned to Cuba in February 1959 as a stringer for Associated Press. This time he learned some disillusioning lessons in Castro's post-revolutionary state. He began to see that Castro had betrayed the revolution and had been a communist all along. He wound up his time in Cuba with a festering hatred of communism and a new direction for his master's thesis: "The Communist Seizure of the Free Cuban Labor Movement."

Given his insights, Brown became entangled with the anti-Castro Cuban exile movement in Miami, rotating between there and Boulder, still working on his college thesis and free-lancing articles whenever and wherever he could. A penchant for writing led to his first tangle with the CIA.

In August 1962, the Denver Post printed an analytical piece written by the young political maverick from Boulder. In that analysis, Brown chronicled specific instances of CIA bungling in Cuba. A week

Right: *SOF Publisher Bob Brown holds a shop-made custom rifle in Pakistan. Gun dealers in the town of Darra Adam Khel produce exotic weapons using only primitive forging methods.*

after publication, Mort Stern, then editorial page editor of the Post, called Brown to relate a conversation that he had had with a well-known Washington columnist, Charles Bartlett. Bartlett told Stern that Brown's article had ended up on President Kennedy's desk. When he had digested the information, JFK called in the CIA brass for an ass-chewing. "If Bob Brown didn't already have a CIA file, he sure got one then." Brown says, laughing at the recollection.

In Cuba he made the aquaintance of Gen. Alberto Baya, author of a primer on guerrilla warfare. In 1960, he obtained a copy and over the next couple years had it translated into English. After writing the introduction the book became known as "150 Questions for a Guerrilla." Although he was convinced of its marketing potential he couldn't find a publisher to accept it.

So Brown and a friend named Bill Jones got together to form Panther Publications in 1963. Brown had the company's first book to publish and Jones had $400. One of its first headquarters was in a drafty mining cabin rented for $15 a month in Wall Street, an old ghost town west of Boulder. Brown remembers many days at work when it was well into the afternoon before it was warm enough to take off his gloves.

After an initial success with Bayo's book—*sans* any royalty payments to the author ("Screw that old commie bastard and his royalties.")—Panther began reprinting and selling Army field manuals and other available military publications about guerrilla warfare and other lurid subjects. While his company struggled on the brink of bankruptcy for 10 years, Brown used meager profits to commute and keep himself on the edge of the action.

Left: *Brown was the first journalist to test fire new Russian AK-74 assault rifle in Darra, Pakistan, September, 1980.*

Reserve training interrupted his post-graduate studies in 1964 and Brown found himself back in uniform. After completing the Infantry Advanced Course and Airborne School at Fort Benning, Brown returned to Miami to keep tabs on the soldier of fortune set. He decided to link up with an operation aimed at spiriting refugees out of Cuba. South Florida at that time abounded in conspiracies for a Bay-of-Pigs-style invasion of Cuba.

The Coast Guard routinely stopped and detained boats cruising off the coast and carrying enough fuel to reach Cuba. Brown and two *gringo* friends thought they could avoid the situation by taking a beat-up cabin cruiser to Key Largo, where they would rendezvous with Cubans who were bringing a cargo of fuel and weapons down from Miami in a car.

Hanging out and drinking in a Key Largo bar which served as the set for many scenes in the Humphrey Bogart film of the same name, Brown and his friends waited for the Cuban connection. Finally Brown took a walk and spotted a parked car occupied by a Cuban he supposed might be their contact. Brown walked up, jerked open the car door

and asked, "Do you have the guns?" They guy inside the car was interrupted in feverish jerking of something besides a car door. Brown slammed the door and walked away in disgust.

"I took it as a sign," he says. "I backed out right then. The Cubans finally showed up and they all took off for Havana. Their boat broke down and if it hadn't been for two ponchos I gave them, which they jury-rigged as sails, they might never have gotten back."

By 1965, Vietnam was beginning to look like a lucrative endeavor for adventurers, even if most of them were wearing uniforms. Brown decided that Southeast Asia should be his next AO. He got a lead on a job with the U.S. Agency for International Development while he was at Fort Bragg for reserve training in the Special Forces Officer Course. His USAID contact in Washington put him in touch with Sam Simpson, the USAID recruiter for Vietnam. Their discussions led to a State Department-paid trip from Denver to D.C., where Brown sat still long enough for interviews and tests in January 1966.

After he returned to Boulder, Brown got a call from Simpson saying the chances were good for a job. They discussed a three-year contract at $15,000 per year, with an additional 25-percent hardship duty allowance. Brown was notified on 7 April 1966 that he had the job pending approval of his security clearance. A week later he was advised that "too many applicants with superior qualifications preclude your selection." Brown remains certain that his Cuba activities were to blame for his failure to get the USAID position.

Still determined to go to Vietnam, Brown requested Army active-duty status. He was still struggling to keep the ailing Panther Publications afloat while researching and writing a book documenting some 30 separate covert CIA operations in southern Florida aimed at toppling Castro. In his yet-unpublished manuscript, Brown applauds the CIA's goals but voices outrage over CIA ineptitude in achieving them.

While waiting for word on his request to get back into uniform, he returned to Miami in November 1966 and joined up as a "peripheral observer" with a group of Haitians, Cubans and Americans attempting to overthrow the right-wing dictatorship in Haiti. It was a failure and some of his merc buddies wound up in legal trouble.

In early 1967, Brown and other Miami buddies planned to highjack a Cuban fishing boat in order to exchange the boat and crewmen for two anti-Castro Cubans who had been jailed in Havana. The mission was subsequently aborted but that didn't bother Brown. He'd been accepted for active service.

Bob Brown and friend strike a pose in front of the team house at Tong Le Chon.

The Army officially recalled Brown as a captain on 9 December 1967. He was issued orders for Fort Bragg, where he entered training for the 5th Special Forces Group. Upon graduation, he was assigned to the 5th Group's G-2 staff with orders pending for Vietnam.

"I had planned on saying nothing about my suspected problem with my security clearance," Brown recalls. "I knew it took six months to complete an investigation and figured if they wanted to send me home after six months, then screw 'em. But I did feel like I should inform my CO of my controversial background since I was assigned to such a sensitive position. When they yanked my file, I was told I could not remain in SF."

Another element of mystery was added when Brown—who presumed he was being punished for some past indiscretion—was offered any school or post he wished.

"I told them that all I wanted was to go to Vietnam," he recalls. Instead, he was assigned to the G-3 staff of Bragg's XVIII Airborne Corps and given command of the Advanced Marksmanship Training Unit.

By 7 May 1968, bad-boy Brown had managed to win a letter of commendation from Lieutenant General Robert H. York, then commanding general of the XVIII Airborne Corps. "It was most gratifying to learn that [Brown's] team won 33 of a possible 56 awards during competition at Camp Blanding, Fla., on 30-31 March 1968," the citation said.

Brown arrived in Vietnam before 1968 ended and was assigned to the 2nd Battalion, 18th Regiment of the 1st Infantry Division. Despite the problems with his security clearance that got him kicked out of Special Forces only months before, the unit decided he should serve as their new battalion intelligence officer. Out of curiosity, Capt. Robert K. Brown, S-2, periodically checked with the Brigade Two-Shop regarding his clearance. He was repeatedly told he had an "interim clearance." A more permanent clearance remained in limbo throughout Brown's 14 months in-country.

Following the 1968 Tet Offensive, many regular U.S. Army battalions were positioned in a static defensive AO on the tactical avenues of approach around Saigon. In this tactical scenario, Brown's task of providing his CO with advice on weather,

Robert K. Brown holding .338 Magnum Winchester and ART scope with Lt. Col. Cruz in El Salvador, 1984.

terrain and the enemy wasn't exactly exciting. He worked his bolt to make contact with the district CIA Phoenix Program and offered whatever help he could provide. This help was in the simple form of getting his battalion CO to provide a company of troops to assist the local Phoenix agents in ambushing large columns of Viet Cong.

Once again Brown had emerged from a fall into the latrine smelling like a rose. He won kudos for doing work that his assignment indicated he officially should not be doing.

On 16 January 1969, James K. Damron, the CIA's province coordinator for the project, wrote Brown a letter of appreciation which cited "outstanding contribution to the Phoenix Program in Gia Dinh Province." That citation lauded Brown for "planning and executing many successful operations against the Viet Cong in Thu Duc" and an attitude and performance which "has gained you respect and admiration from personnel associated with the Phoenix Program."

That was nice, but Brown still wanted to serve with Special Forces. He use the CIA commendation as a lever to gain an interview with an old acquaintance and fellow Coloradan, Lieutenant Colonel John Paul Vann, a retiree who was then in Vietnam working for the State Department. Brown told Vann he was "tired of being a leg" and wanted to get back in SF. Vann wrote a friend of his, Colonel Harold R. Aaron, who happened to be CO of the 5th SFG, and within a week Brown had orders to take over as CO of team A-334 at Tong LeChon. Vann wrote Aaron that Brown "is of particular interest. . . . He is one of our leading experts on counterinsurgency. . . . Not covered in [Brown's] resumé was a period of activity related to the Cuban affair which was rather interesting."

Bob still gets angry when he talks about his six months as an A-Team leader. He is convinced that his ARVN counterpart in the camp, a Captain Long, and the ARVN B-Team CO, Major Long (no relation) were conspiring with the Viet Cong and American soldiers to profiteer on the black market. Brown determined that his regular ration shipments for Montagnard Strikers were 10 percent short of specifications. And Capt. Long was selling the supplemental rations at highly inflated prices. But Brown stopped the supplemental shipments, thereby "breaking Long's rice bowl," or derailing the Vietnamese officer's scam.

Refusing to stand still for the rip-off, Brown complained to everyone in the chain of command who would listen. It reflected on his officer's efficiency reports, but Brown credits the experience as valuable in later life when he was struggling against similar odds to keep SOF going.

"Capt. Brown had trouble establishing rapport and an effective working relationship with his [ARVN] counterpart, the Vietnamese Special Forces Camp Commander," wrote Lieutenant Colonel Charles B. Cox. "As he explained to me, he did not trust the camp commander and was convinced that he was disloyal and involved in graft. Capt. Brown was counseled on the necessity for maintaining good working relations with his counterpart, but not much progress was made."

That's not the way Brown sees it. He considers it progress that Capt. Long was later identified as a Viet Cong collaborator and shot. Before that happened the turncoat ARVN officer managed to involve Brown in an errand that led to an ambush and 14 frag wounds from an exploding 82mm mortar round. Brown recalls that he woke from surgery and hobbled into his C-Team leader's office yelling about his in-country counterpart being a VC. Nothing was done until Capt. Long tipped his hand.

Completing his Vietnam tour as a political warfare officer in Nha Trang, Brown returned to the U.S. to find himself ejected a second time from Special Forces. He was assigned as CO of an Army Reserve Basic Training Company at Fort Leonard Wood, Mo., and that led to his eventual release from active duty on 30 April 1970, exactly five years to the day before Saigon and all of South Vietnam fell to the communists. He remained active in the Army Reserve and eventually retired as a lieutenant colonel in January 1985.

Brown returned to Boulder and found his pet Panther Publications in trouble. Brown bought out Bill Jones' interest in Panther Publications for $1,000. He then sold half-interest in the fledgling publishing company for $5,000 to an old Florida buddy, Peder Lund, who had also been a Special Forces team leader in Vietnam. Brown and Lund renamed the company Paladin Press and continued to reprint hard-to-get military manuals.

For the next four years Brown supplemented his income with odd jobs, including construction laborer, private investigator and instructor at the Boulder Athletic Club, where he spent some nights sleeping on the massage table. Paladin Press, like Brown, was scraping by on a thin margin. He had completed his Master's degree, done more work on his CIA-Cuba book and even contemplated an academic career. His application for a Ph.D. program in Boulder was voted down 3-2 by the CU staff. Brown says one professor later told him the rejection was based on his outspoken, anti-communist, conservative political views.

The man who'd spray-painted "Viva Castro" on the chemistry building had come full circle. It was time to strike off in a new direction.

In 1974, Brown made Lund a buy-or-sell offer for Paladin Press. Lund bought Brown's interest for

$15,000. Hoping to finish his book on the CIA and make a side-trip to an interesting little hassle in a place called Rhodesia, Brown accepted the cash and hit the road, headed for Spain.

In Madrid, Brown hooked up with Mike Acoca, a former *LIFE* magazine writer then reporting for *Newsweek*. Acoca and Brown were friends from Cuba days and they planned to collaborate on Brown's book. True to what was becoming fate for Brown, Lisbon erupted into civil violence on 24 April 1974, and Acoca had to go cover it. Brown tagged along for a couple of weeks, but grew tired of the street marches and random shooting. He took off for Rhodesia and a more intriguing situation.

Prior to leaving the U.S., Brown had tried to sell the idea for a story on mercenaries in Rhodesia to America's three most prominent men's adventure magazines: *True, Argosy* and *Saga*. Two turned him down and the other never answered Brown's query. He decided he could free-lance the story when it was complete. While in Rhodesia, an American merc told Brown of several compatriots who were joining the Sultan of Oman to suppress a leftist insurgency. Brown declined an invitation to join them, but kept the information on the Sultan of Oman's recruitment for future reference.

Once back in Colorado, Brown immediately set out to do two things. One, he wanted to defray expenses from his trip, so he again tried to sell his piece on U.S. soldiers of fortune in Rhodesia. One editor told Brown "we're trying to get away from that hairy-chested stuff."

"Hell," said Brown, recalling the incident on the 10th Anniversary of *Soldier of Fortune*. "I thought hairy-chested was the name of the game for this type of magazine." *True* reconsidered and paid Brown $750 for his article, but the magazine went bankrupt before his story was published.

Brown's other priority on returning to Boulder was to capitalize on the recruiting of mercs for service to the Sultan of Oman. Bob wrote the Omani Defense Minister and soon received a 40-page information packet that included terms of employment. He mimeographed the information and began advertising the recruitment packet in *Shotgun News* and similar publications. Brown got a respectable response, he said, until *Newsweek* reprinted his ad as part of a report on the recruiting of mercs in the U.S. The postal floodgates opened and Brown eventually made a tidy $5,000 profit.

The response to that advertisement and the financial collapse of the leading men's adventure magazines in the early 1970s convinced Brown there was a void in the publishing marketplace that needed to be filled. He was also convinced there was a crying need for someone to herald the sacrifice and profes-

sionalism of America's returning Vietnam Veterans.

A linchpin in the survival of *Soldier of Fortune* was that Brown knew the Vietnam Veterans never received due recognition upon their return. Brown's unbending position remains that Vietnam provided just as many heroes as did both world wars and Korea. "But how many Sergeant York or Audie Murphy stories did the news media give the U.S. public from Vietnam?" asked Brown, "None. My magazine has corrected that."

In late 1974, using his $5,000 from the Oman ad and the $750 from *True*, Brown began putting together a promotional brochure touting the concept of a magazine about mercenaries and professional adventurers to be called *Soldier of Fortune*. He obtained additional funds from Colonel Alex McColl, another Army reservist who is SOF's Director of Special Projects, and Don McLean. Brown bought out their interests after a year.

He bought a mailing list from a gun magazine, sent out the promotional brochures in February 1975 and waited. He had already decided he needed a $36,000 budget to keep the magazine afloat for one year and that he'd have to make that on the first issue. More than 4,000 Americans—mostly Vietnam Veterans—sent him $8 for a one-year subscription to Brown's startling new magazine. Ad revenues put Brown over the top of his bottom budget line, but he was taking no chances on the possibility of a last-minute fold. For two months he rode around with subscription checks in a cigar box under the front seat of his GTO. He figured it would just be easier to simply mail the uncashed checks back if his heretical venture went belly-up.

The first issue of SOF—which carried Brown's story on U.S. mercs in Rhodesia and a now-famous, grisly photograph of a terrorist victim with the top of his head shot off—was mailed to subscribers in July 1975.

SOF's Wide World of War

South West Africa

Bob Brown, SOF Publisher, visited many forward operational bases near the Angolan border during 1978. His notes and observations that follow provide an excellent overview of the situation in South West Africa during those days of struggle.

Terrain and Situation

The border between Angola and South West Africa in the military area I visited runs for about 450 kilometers. The terrain, stark, flat, and featureless, is covered with 20-foot-high bush.

When questioned as to how the South Africans secured such a long porous border, (the whole border is essentially one big "avenue of approach") one SWA base commander replied, "I concentrate most of my operations in and around the more densely populated areas. Of course, we also run ops along the border on an irregular basis to preclude establishing any definite pattern. I also deploy my units in depth."

The border area is broken down into two "Military Areas," which have a troop strength the equivalent of a U.S. Army reinforced brigade. The number of companies in each battalion is determined by that particular battalion's mission and the number of square kilometers in its particular area of operations. The same criteria are used to determine the number of armored vehicles assigned to each

Above: *S.A. armored cars on patrol.*

battalion, which in turn are attached to the battalion's companies.

A permanent mobile reserve of paratroopers are stationed at the Military Area Headquarters. Normally, they are inserted by helicopter.

The importance of choppers in this nasty terrorist war is as great as in Vietnam. All untarred roads must be carefully swept for mines, which makes it impossible to reinforce quickly by land. Furthermore, those roads become impassible during the three-month rainy season.

The battalion and company bases are fairly permanent and can be compared to our fire bases in Vietnam. Twenty to 25 percent of the combat troops remain at the bases; the remainder are out on ops.

Above: *Ovambo store owner poses by his pickup, blown up by SWAPO terrorists.*

Below, right: *Helicopter squadron returning to Angola base.*

Below, left: *Members of S.A. parachute battalion head for rifle range. Note unique silhouette targets.*

Above: *Captured SWAPO terrorist.*
Right: *Dead SWAPO terrorist.*
Left: *Taking a break during patrol in hot, dusty Namibian Desert near Walvis Bay.*

Terrorist Modus Operandi

SWAPO's base camps are located between 30 to 100 kilometers north of the Angolan border. These bases serve as SWAPO's section headquarters. Africans who are kidnapped or recruited or lured to SWAPO's standards under false pretenses rendezvous on these bases where they are evaluated. The most qualified are sent overseas for specialist or advanced training. The remainder are trained at these bases for three to six months.

As one South African Army colonel pointed out, "This is not a thorough training cycle when you consider that all instruction must be conducted with interpreters."

After completing training, the SWAPO recruit moves to an operational base which normally is close by a logistical base. Subsequently, they are sent to operational bases six to 10 kilometers inside Angola, which are temporary and which are moved on a weekly or even daily basis.

From these bases they infiltrate into South West Africa. Within 10 klicks of the border SWAPO units tend to be aggressive. They will pick the time and place if they wish to engage. Normally, they rely on a short but heavy volume of fire and then break contact. In this type of contact the MGs will fire a belt of ammo, the riflemen a couple of magazines of AK ammo and the RPG team will fire one or two rounds during a one or two minute contact.

However, as SWAPO units move further south into South West Africa or Namibia, as it is called, they avoid contact with the security forces. In fact, the only contacts are by accident, unless the security forces are successful in tracking them down.

Above: *S.A. recruits receive patrol briefing before night operations.*

52

Right: *White volunteer soldiers. Many Portuguese mercenaries were often seen at the head of FNLA regiments.*

Far right: *Dog patrol in arid bush country of South West Africa.*

Above: *Black Ovambo soldier on parade.*

Left: *White officer attached to the unit explains patrol operations.*

Terrorist Intimidation

The terrorists' main objective is to intimidate the Ovambo population in South West Africa in order to win the elections and take total control. The type of intimidation differs from the areas near the SWA-Angola border and further south. Along the border murder and atrocities are common practice, while further south threats are more common. The terrs will tell tribesmen, "Either you do as we say or the same thing will happen to you that happened to 'Mr. X' up near the border."

The main terrorist objective is to weaken the influence of the local tribal authorities. SWAPO labels the present Ovambo leadership as puppets of the South Africans, calling them "black Boers."

SWAPO's policy of intimidation got into full swing in 1976. It was and is targeted against tribal authorities, the Ovambo government, and those supporting the government. All churches and missionaries are also targeted, with the exception of the Lutheran church, which is well disposed toward and aids SWAPO.

Major Hans Stempfle, who was my escort during my tour of the SWA operational area, explained why mutilation was a far more effective intimidation tool than murder. "The Ovambo tribesman is very stoic and fatalistic. Simple killing does not have the impact on tribal members as does, say, cutting off arms and legs or noses and ears and then forcing the victim's wife to fry and eat the flesh."

Intelligence

Developing intelligence is difficult. One S.A. Sergeant Major stated, "Who ever gives the last political speech or shows a presence in a *kraal* is the party who has current influence in the area. If a SWAPO representative speaks, the tribesman will appear to be an ardent SWAPO supporter and vice-versa when we appear. The natives volunteer information like children. If they feel safe, they are just as likely to provide us with intelligence as they are to SWAPO. If we follow spoor into a *kraal* and there is a terrorist presence, the people will say they have been asleep and have seen nothing."

Another factor affecting the situation is the "strongman concept" amongst the tribes. They are far more impressed with "strength" than they are political ideologies. For instance, after South African forces struck SWAPO bases in Angola on 31 May 1978, the "Jungle Telegraph" quickly passed the word throughout the tribes. For almost two months, the tribesmen were far more cooperative and friendly to the South African forces and provided a great deal more intelligence than normal.

Left: *Major Willie Snyders explains a tactic to section leader. In the bush, hand signals are used.*

Above: *As is typical in communist countries, youth are used in war.*

Left: *UNITA and FNLA soldiers celebrating victory in 1975.*

Above: *Dr. Jonas Savimbi, charismatic leader of UNITA.*

Right: *Mounted S.A. unit tracks terrorist base camp.*

Population and Resources Control

As in any insurgency situation, countermeasures, to be effective, must be tailored to the particular culture threatened by the insurgents. A case in point is the concept of "Protected Villages" as they are known in Rhodesia or "Strategic Hamlets" as they were known in Vietnam.

The South African authorities considered this option but decided that due to the Ovambo tribal culture and traditions, such a program would cause more trouble than it would be worth; that the hostility such a program would engender would outweigh the benefits of separating the bulk of the tribesmen from the terrorists.

"We try to separate the terrorists from the local population," commented a South African Army colonel, "but it is extremely difficult as the terrorists are members of the same tribe and often have family in the *kraals* they enter. The terrorists wear civilian clothes underneath their camouflage uniforms and simply strip them off and blend in with the populace when necessary."

Safe Areas

As has been the case in past counter-insurgent programs, the individual South African soldier feels he could be more effective if the South African forces could follow one of the cardinal principles of counter-insurgency—strike the enemy in his base camps or safe areas.

"We can't do it as regularly as we would like," pointed out one South African officer who had spent two years in the operational area, "due to the political and international situation. You saw what happened when we went in and hit the SWAPO bases in Angola." Shades of Vietnam!

Left: *Youngsters start learning art of war at early age.*

S.A. Armor

The South Africans, like the Rhodesians, have developed a number of unique, mine-proofed wheeled armored vehicles that operate effectively in the flat, sandy, bush-covered terrain.

One of the most impressive of this generation of infantry fighting vehicles is the "Ratel," named after a South African mammal similar to our badger. Development of the Ratel began three years ago and the first one rolled off the production line only a year later. It appears there is a lesson here that could be learned by foot-dragging boondogglers in our own U.S. Army procurement system, who are still diddling around (and have been for several years) trying to develop and manufacture a replacement for the now obsolescent M-113 armored personnel carrier.

The Ratel carries a 20mm main gun (type is classified) and a 7.62mm coaxally mounted machine gun; weighs 3500 pounds and has a maximum speed of 65 mph. Range, type of suspension, armor and engine are classified. It carries a crew of 10. The crew commander is also the driver, operates the 20mm gun, and leads the crew when they dismount. According to the 1978 edition of *Defense and Foreign Affairs Handbook*, the South Africans have 500 Ratels available.

The South Africans have a number of more recently developed armored troop carriers in their inventory in South West Africa, which are operational though also classified. These armored troop carriers have been specifically designed to provide protection from the most powerful of Russian-made landmines.

The bottoms of the bodies or hulls of these vehicles are V-shaped, with the bottom of the "V" facing the ground. This type of configuration deflects blast

Left: Biker in South Africa revs up to fight terrorists.

and heat effects from an exploding landmine. A compartment in the bottom of the V-shaped hull is filled with water—also to absorb blast and to serve as a blast coolant. All seats are fitted with helicopter-type seat belts. Passengers who have not "buckled up" have been thrown 30 yards when their vehicle has hit a landmine.

The South Africans have also developed a number of Star-Trek-looking mine-clearing vehicles with such colorful nicknames as the "Praying Mantis" and the "Grasshopper." I had a chance to observe these vehicles but they also are classified and so no photographs or specifications.

Since the introduction of these new mine-proofed vehicles, six of them have hit landmines but there have been no casualties.

Far left: *South African forces rely on locals for information and their bikes for cover when operating in SWAPO areas.*

Left: *Close-up of Bushman corporal in dress uniform.*

Mounted Infantry in a COIN Role

I had hoped to ride with the South African mounted infantry. Unfortunately, by the time we choppered into their base, they had been called out on an operation. This war, like all others, waits for no one, not even SOF.

As dusk began to fall, the small troop of horsemen returned. After grooming, feeding and watering their horses, 2nd Lieutenant Pieter Alberts, CO of the mounted infantry section assigned to this particular operational base, joined myself and other S.A. troopers for a barbecue where we discussed the merits of horses in combat.

Lt. Alberts entered the army in January 1977 and after completing the "Junior Officer's Course" was commissioned. As he had an equestrian background—his parents manage a resort and stable—he was assigned to the SA Army's Equestrian Center, for a six month course. After completing said course, he spent several weeks receiving intensive CI training at a forward training base in South

West Africa before moving to his present assignment.

All equestrian trainees receive the same basic training with horses no matter what their background has been or what their skill level is.

And basic means just that. A trainee first learns to bridle, saddle, mount, and groom his horse. Concurrently, he learns how to "maintain" or care for his horse in the field and how to get the most out of the animal. Each man also learns the rudiments of bush medicine, as it is seldom that the troop is within call of a vet when on ops.

"The advantages of horses are they are mobile and fast. Also, you have better visibility from the back of a horse." Alberts continued, "As the old saying goes, 'The infantry fights upward, the cavalry downward.' "

As the trainee becomes accustomed to riding, he is taught PT while mounted and learns to jump a variety of obstacles—ditches, fences, brush, etc. He learns to control and direct his horse with his thighs, leaving his hands free to operate his side-

Right: *Taking a river on the run.*

Below: *Trackers at work.*

arm. I tried riding on a ranch in Rhodesia with a FN in one hand and reins in the other. It's difficult at best. At a trot or gallop, well . . .

"Every horse has different characteristics and responds differently in various situations," Alberts commented. "To obtain maximum efficiency when on operations, all trainees keep the same mount through the six months' training period and during their tour in the operational area. In other words, the trooper and horse remain as a team up to 15 months, assuming neither are injured.

An example of the use of the speed of the horse was demonstrated when one mounted unit cut tracks of a band of terrs at dawn that was already 24 hours old. Before nightfall, the mounted infantry had caught up with and engaged the fleeing terrorists, inflicting several casualties.

"Normally," Alberts continued, "we do everything an infantry man does, only on horseback. Teamwork is imperative. You must be able to anticipate what your mates are going to do.

"The section leader is supposed to make the decision regarding counter ambush response, but many times we react instinctively. Once dismounted the well trained horse will move out of the danger area and wait for his rider. We normally only use semi-auto fire from our FNs on horseback."

The major disadvantage of using horses is that they are living creatures; they do get sick. And it takes five or six months for a horse/rider team to obtain maximum efficiency.

As SOF staffer Venter mentioned in an article describing ops in S.W.A., motorbikes have been successfully incorporated in the COIN operations. The 250 cc scrambler type bikes are used to pursue terrs when they are on the run and/or have a long lead. They are also used for reconnaissance. Trackers, who follow terr spoor, can read spoor while riding their bikes. One exceptional tracker can read a very plain set of tracks while zipping along at nearly 40 mph!

When chasing down terrs, the bikers will move ahead to establish contact. Riders and saddled horses, ready for immediate deployment, follow in vans pulled by four-wheel drive trucks.

"You must remember that our horses are not used in actual combat unless it can not be helped. They are used primarily as a means of transportation. If, however, a horse section is ambushed, the immediate reaction drill procedure calls for us to charge the ambush position—if the range is 20 to 30 meters. We then dismount and move back to engage the enemy. If the ambush is triggered at a greater range, the unit dismounts to engage or may attempt to flank the enemy on horseback. We are also used on follow-up operations and as a blocking force.

Above: *Two members of the crack Selous Scouts with captured weapons.*

Right: *Another aspect of interior security, the bicycle brigade.*

SOF'S Rhodesian Fire Fight

SOF made quite a reputation in the early years of publication for fearless, firsthand reporting from the bloody battlefields of Rhodesia. Their efforts in that ill-fated African nation and their support of the Rhodesian government in operations against communist insurgents gained them two unfortunate, undeserved labels: racists and mercenaries. They are neither. On the other hand, SOF has never avoided consorting with genuine mercs to insure readers get the look and feel of Third World battlefields. What follows is an abbreviated version of a story from August 1980 by Joe Tragger which covers the Rhodesian front.

When you're in the Rhodesian bush, you go to bed shortly after the sun goes down and get up before it breaks the morning sky. The night before an operation you may not sleep well, wondering if you checked everything and what tomorrow will bring.

As the sun comes up, it's time to load the seven-fives (armored vehicles) and ride to the operations area. The 14 members of the stick kid around as all soldiers do before an operation. The machine-gunner on the stick tells us "the Major" has found terrs on more than 70 percent of his operations. They say "the Major" is good luck.

He is Darrell Winkler, former Officer Commanding the Rhodesian Armored Regiment, now OC Rhodesian African Rifles.

Other members of the stick are Jerry O'Brian, Great Britain, ex-French Foreign Legionnaire; Michael (Reb) Pierce, American machine-gunner; Yves Devay, veteran of the Belgian Army; "the Mechanic," the only white Rhodesian native on the operation; and SOF staff members, Editor/Publisher Robert K. Brown, Art Director Craig Nunn, Associate Editor N. E. MacDougald and myself, Joe Tragger. The remaining members of the stick are all black troopies of the RAR, a fine unit.

A few kilometers out from the base camp, all kidding stops. Smiles disappear as tension grows; faces tighten. You put a round in the chamber and start watching the bush. You're not too concerned about mines—Rhodesian seven-fives are mine-proof except from the larger Soviet anti-tank mines.

Why worry? Hell, nothing can be done about that.

We're in one of the Tribal Trust Lands (TTLs). Silobela, it is called. There are about 70 or so terrs in this area, according to intelligence sources.

About 12 to 14 clicks from the base camp we stop. Security is posted around the seven-fives and last-minute instructions given. Drivers are given pick-up points, watches are synchronized and we're off.

We are near one of the branches of the Gwelo River. We head southwest, hoping our intel is good. The pace is quick with little noise and everyone is alert. Tomorrow will bring election returns—and maybe peace. No one wants to be the last killed.

In each *kraal* (village) the RAR sergeant questions the locals about the terrs in relaxed and easy exchanges. We are given information about where the terrs camped overnight. The major decides to split the stick. His section will delay and head straight for the terr camp. We are to cut a big arc behind and set up an ambush as the terrs are driven into us.

The terrs prefer to use the many rivers of this area for guidance so we set a quick pace to the Damba River, a branch of the Gwelo. A little over two kilometers from the branch is Damba Dip, where we hope to catch them in ambush. Our quick pace continues as we move on.

Arriving at the *kraal* the story is the same. Yes, terrs are in the area now. A few locals say the last time there were terrs around was 23 November of last year. Funny how that date sticks in their minds, as if someone has programmed them. The fear in their eyes shows the Popular Front has gotten to them.

Another five kilometers or so and we take a break about 50 meters from the banks of the Toto-lolo River. A couple of minutes for water, biscuits and jam, shift your load and cross the river. About a hundred meters on the far bank we spot a leopard, stalking a large bird. I feel better about our movement, as we haven't disturbed the big cat. We're moving quick and easy now—and then we hear it.

Reb and his Mag 58, a short burst, then rifle grenades, AK-47s and the 870 shotgun Craig is carrying. The contact is about one kilometer north of us. We cover about 700 meters in a dead run and halt to duck rounds coming through the foliage. Devay wants to charge into the contact, but we have no radio and the major won't know from which direction we're "coming in." We hear sporadic firing—then movement—they're chasing the terrs. We move into a blocking position and wait. Just like any other war, there's a lot of waiting in Rhodesia.

When no terrs show up, we move off to our RP (rendezvous point) at the Do Me Good Store. When the rest of the stick joins us, Bob Brown is clearly elated. In all his trips to Rhodesia, this is his first operation leading to a fire fight.

His story is typical of a Rhodesian bush contact:

"As the first AK rounds cracked over our heads I came to the conclusion that corn stubble makes lousy cover. Reb was on my right and he triggered some short bursts from the MAG. I wondered whether the bastards would fight or shoot and run as usual.

"Blam! Blam! Two terr rifle grenades exploded on line about 10 meters away from the MAG position. They had the range but were off-target. Major Winkler yelled for cover and ordered me with him around the flank. We blasted the bush with our Mini-14s and made for a position where we could cover Reb. He advanced the MAG another 30 meters under our fire. Then the gun jammed. We couldn't clear it so I broke out the camera to take a few pictures.

"We continued the sweep looking for spoor, movement or reflection from an AK. Then they tossed another rifle grenade about 10 meters to my rear. It was close as hell."

It was also inconclusive like so many terr contacts in this brutal bush war. We returned to our RAR

Right: *Machine gun instruction at a regimental battle camp.*

Above: *Alouette chopper in Rhodesian service.*

Below, right: *Off for a bit of country duty with an FN.*

Left: *Para "kitting up" for a jump. Machine gun will be slung over his shoulder just like a rifle.*

base camp to await the election results.

We slept well but awoke to bad news for Rhodesia. It's official. Mugabe has won. Bishop Muzerewa against Mugabe, God vs. the AK-47. Take heed, Jimmy Carter, the AK won. The AK is a decisive campaign manager.

In Salisbury, we hear rumors of plans to burn the city, of a hit list Mugabe's people have that includes all SOF staff members. Back at the RAR base camp, Rhodesia's bush-weary professional soldiers cannot believe the army will no longer fight. We hear of PF plans to remove all members of the Rhodesian military who are not Rhodesian citizens. What I did not realize before coming to Rhodesia is how many Brits, Aussies, Yanks, Irish, Belgians, Kiwis and many others are in the Rhodesian army. If they leave, the security gap cannot be filled.

We get a message that it's time for us to get back to America, England, Belgium—anywhere but here. Nobody knows what Mugabe will do. Very few trust him. Communist or not, he was the enemy. I wish Zimbabwe-Rhodesia well. I wish Mugabe well. But my best wishes are all he will get—not me. Hell, it was a Washington/London/Moscow war anyway. They are all politician's wars, but no one imagined there were so many politicians in Salisbury.

Uganda

In October of 1979, SOF Staffer, Al Venter, gave this on-the-spot account of the situation in newly-liberated Uganda.

Tanzania has received a mixed reception in newly-liberated Uganda. Although largely Tanzanian troops have helped the Ugandan rebels overthrow the tyrant Amin, they have left much desolation in their wake. In the words of a Swedish journalist, a member of our team: "Amin put Uganda back 20 years; the Tanzanians have doubled that figure."

Looting has been widespread. Much of it took place immediately after the invasion, initiated by drunken mobs of Tanzanian troops. Anything that could be moved was placed on empty ammunition trucks and sent back to Dar es Salaam. Looting has been so bad that almost every stove and refrigerator in the Ugandan capital and Jinja have been sent back to Tanzania.

Looting also extended to ordinary household goods. One young Tanzanian lieutenant had 14 wrist-watches in a bag and another five on his arm. He was proud of his acquisitions. Did he get them from Amin's soldiers?

"No, we take from the people. They must pay for this war."

Even in the hotel, Tanzanian troops had removed door handles, light-switches and tap faucets to take "back home."

The war itself has not been nearly as intense as the Biafran campaign, or as widespread as Ethiopia's vicious Ogaden struggle. Compared to Rhodesia, it's a smoke break. What is important is that one African state has used its army to overthrow another—an important precedent—and 25,000 men-in-arms were used in the campaign.

The war has lasted longer than anticipated with the invasion force vanguard bludgeoning its way through Amin's combined forces at Mutukula on the Uganda-Tanzania border on January 22 this year. Mopping-up will continue for some time to come.

Although hostilities occurred before then, with conflict see-sawing across the border between the two nations, Mutukula was the first decisive Tanzanian blow. Amin's forces reeled in disarray and fled.

Lethargy has characterized operations on the part of both forces. Granted, there have been a few humdinger battles, such as the fight for Mbarara, south of Kampala, and another at Lira where Tanzania is reported to have lost more than 200 men. Otherwise, the war was a slow northward grind with the majority of Amin's forces getting out while the going was good.

Above: *Victorious Tanzanian at a border checkpoint.*

Left: *A dead Lybian.*

Below: *Moments after this Ugandan was questioned, he was executed. Guards are wielding both AKs and SKSs.*

Afghanistan: More than the Bear Bargained For

SOF Correspondents were not the first Westerners to cover the situation in Afghanistan after the Soviet invasion prompted formation of the rugged mujahideen resistance movement but some of the most spectacular—and widely read—reporting appeared in the pages of this magazine. SOF made friends among the Freedom Fighters and they were allowed to gain close-up looks at the fighting as well as valuable, exclusive reports, on Soviet weaponry including the AK-74 assault rifle, the BG-15 and AGS-17 grenade launchers, the RPG-18 and the AKR "Krinkov." But the greatest feat in Afghanistan was bringing the plight of the resistance fighters and the ruthlessness of the Moscow-Kabul tyrants to the attention of the American public as this story from the October 1980 issue, written by Galen L. Geer, illustrates.

I looked down from my precarious perch on the camel at the lone mujahideen walking beside me. For the past hour the Russians had been bombing the valley across the mountain from us and he shuddered with each new wave of explosions. When we took a break, I pulled the interpreter, Abdul Massai, to one side and asked him why one man seemed to be so affected by the bombings while the others were indifferent to it.

"Because the bombs are falling on his home," Massai said, then walked away. His matter-of-fact tone confused me and I struggled to crawl back onto the camel. When I was finally settled, I lit a cigarette and thought about the visions which must have been going through the Afghan's mind as he heard the bombs fall on the only home he had known since birth.

Mortar round impacts near Russian fort under siege. Only 11 rounds were fired this day — more than a week's supply.

Insert: One of the many PFM1s (air-developed anti-personnel mine) scattered across the highland trails from Soviet helicopters.

Above: *Soviet BTR-60BP APCs in Afghanistan. Afghans have discovered that bursts of automatic-weapons fire into the forward tire of the standard Soviet APC will not only deflate it, but at least one round will penetrate the mild-steel wheel well and kill the driver.*

We spent that night in another one of the countless mud huts the mujahideen use as "safe houses" throughout Afghanistan. After a skimpy meal of rice, tea and bread we settled on the floor to sleep. In the distance the firing continued. Artillery had taken over from Soviet air. Ivan was going all out for something. I wondered if he was successful.

Around 0700 the pounding started again. We moved off and as we covered the desert valleys and rocky hills I kept thinking that MiGs and choppers make for a hell of a one-sided battle. Still the mujahideen seemed to be holding their own in the jihad—the Afghan holy war.

The jihad is a confusing mixture of historical and current Afghan problems. The war's general purpose—both tribal and political factions in Pakistan agree—is to rid Afghanistan of Russians. But each resistance group seems to be going in a different direction and Western observers are left confused and frustrated.

Even without a holy war against the Soviets, the Afghans would be happy to fight them on a hit-or-miss basis because it is good sport. The basis for fighting is centuries old, not a sudden outpouring of national pride. Seeds of this jihad were first spread a decade ago when many of today's political leaders in the mujahideen began to denounce the communists then active in Afghanistan. Since many of them were also spiritual leaders they were able to whip up an anti-communist fever among the people shortly after the coup which led to the first communist regime. That led to the first phase of the present Afghan war.

Below: *Soviet troops move out from a landing zone.*

Left: *SOF staffer Jim Coyne blasts away with 12.7 mm model 38/46 DShK Degtyarev HMG. This was Coyne's second trip to Afghanistan.*

Another important point to understand about the Afghan war is that there is no death for the mujahideen in battle. Because they have become mujahideen, holy warriors, they have already had their Islamic last rites and believe they are dead. When they do die in battle they are accepted in heaven by Mohammed. They become *shaheed* and live forever and their graves become shrines.

The spiritual leaders point out that the Russians cannot defeat the mujahideen, because for every one who is killed in battle, 10 more will rise in his place. It sounds crazy to Westerners until they see the fever pitch of Afghans leaving Pakistan's tribal areas for Afghanistan, and listen to the tales of glory when mujahideen are killed. New recruits swarm from refugee camps around Peshawar to join the resistance when they hear such tales.

Because Afghan people have spent generations fighting in holy wars, local brush wars and national wars, each family, each generation, has its own history of glory. The present jihad, for many of the Afghan men, is a chance to expand that glory. By appealing to their religious devotion, their sense of injustice over the destruction of mosques, the murder of women and children, the bombing of villages, the groups in Peshawar have a bottomless well of manpower. Their only real shortage is weapons.

Eighty to ninety percent of all Afghan freedom fighters are still armed with Enfields; the rest may have a few AKs and SKSs.

To most of us who love a good fight and are willing to jump at nearly any chance to get in a few licks on Ivan, Afghanistan would appear to be the place to do it—until the holy war begins to come into focus. There are not a lot of mercs around who want to get mixed up in a war where last rites are handed out before the battle.

As the war continues, Ivan will tighten his hold on most of the major roads in Afghanistan and force the mujahideen further back into the mountains. It won't make much difference. There is no end to a holy war. It will, most likely, become a PLO-type operation in the next few months unless Western aid begins to filter into the mujahideen camps. Even with the lack of arms, Afghans can keep the Russians from controlling the rugged desert or the lush pine forests of the mountains. Nothing short of a Berlin-type wall is going to seal the border with Pakistan and Iran.

One of my guides on my trip through Paktia Province explained the Jihad and the differences between the resistance groups. "First, we kill the Russians in the holy war—then we start the real war to find our own government."

It is going to be a very long war for Ivan and the Afghan people. There was plenty of time to dwell on that as we panted up the rugged, rocky slopes of this area. My legs felt like lumps of lead. Below, the valley stretched out in an endless sea of brown. A narrow ribbon of blue water and green vegetation along the river's edge were the only signs of life in the Afghan desert. I looked around at the lush pine forest we had climbed to and the half-dozen smiling Afghan mujahideen who seemed unaffected by the past four hours of climbing.

Observing the fall of a mortar round.

Above: *Live Soviet PFM-1 mine.*

Right: *Mujahideen do not have the expertise to repair or utilize captured Russian radios. Their communications are usually limited to a few walky-talkies with a range of about three klicks.*

For 10 days I had been walking through Afghanistan with the mujahideen and had decided they were among the world's best—although weirdest—fighting men. At times they live off little more than dry bread and tea. They wear nothing more than old worn-out sandals, whether climbing mountains or walking through the desert, and don't give a damn about how far it is to the next water hole or if they will get a decent meal. Since one of my assignments for SOF had been to find out how the individual mujahideen lived so we could present readers with a profile of these famous warriors, I observed them closely on our trek. One of the most interesting aspects I found while with the mujahideen was their ability to withstand the demanding nomadic life required by combat in this unforgiving land. They do not carry rations or canteens. They get their water where they find it and meals are often little more than a little rice, *nan*—their dry wheat bread, a staple in the Mideast—and tea. Although they offer to pay for every meal, most are given freely by the mountain people to support the holy war.

A full day's march for the mujahideen begins before dawn. As soon as the morning's prayers are over, they drink a few cups of tea, tear off a few hunks of bread, then gather together their weapons and what little equipment might be carried on camels or donkeys. Then they go and go hard.

Each man carries his own weapon—anything from a World War II Russian pistol to a modern AK-74 captured in recent fighting. Their range of weapons includes shotguns, ancient Chinese machine guns and Enfields.

The most ammunition I found carried by a single man was 50 rounds. One 20-year-old carried a Colt .38 Gold Cup National Match pistol and wore his rounds in a bandoleer across his chest. Most Afghans have from 20 to 30 rounds at any one time. Misfires are saved and put back in the belt to have new primers put in them. Their other standard weapon is an Afghan dagger, a wicked-looking knife with a T-shaped blade and camelbone handle that is curled at the end.

Below: *Mujahid wounded while attempting to deactivate PFM-1 mine by throwing rocks at it.*

One mujahideen irony is their childlike love for bright colors and flowers. All weapons are decorated with colorful beads and leather strips over the stocks and barrels. When walking past a field of flowers, they cannot resist stopping to pick a few to stuff in their weapons, hats or clothing. Their ferocity and gentleness are a paradox. A family whose donkey had sprained a leg was struggling with the animal's load when our group walked past. The mujahideen picked up the family's load and carried it down to the river. They waited for the man and his wife to bring the limping animal along before leaving.

Before going into Afghanistan I heard tales of the man-killing pace of the mujahideen. I discovered their "killer pace" would be a crawl to military types who think of cross-country travel as a route step or march. The mujahideen take small, slow steps in an unchanging rhythm both up and down hills. Where most of us tend to pick up our pace as we go down a trail, the Afghans maintain the same pace, the same distance with each step to conserve energy and moisture in the blazing desert sun. Until I learned to match my steps to theirs I was always either way behind or way in front of the group. Once I figured out what they were doing, I was able to stay with them and live off the meager rations as well as they did.

I found the mujahideen more interesting and determined than any other fighting men I've spent time with in the field. In their simple, unassuming way, with their determination to throw the Russians out, the mujahideen are capturing the world's attention and holding the Russian bear at bay. It might not be a bad idea to send a few of our own NCOs and officers over to take lessons from them.

Left: *Adjusting the bipod on a Stokes-Brandt three-inch mortar. The Stokes-Brandt was first introduced into British service in 1936.*

Right: *Preparing Chinese 82mm mortar-round fuses, primary cartridges and increment charges.*

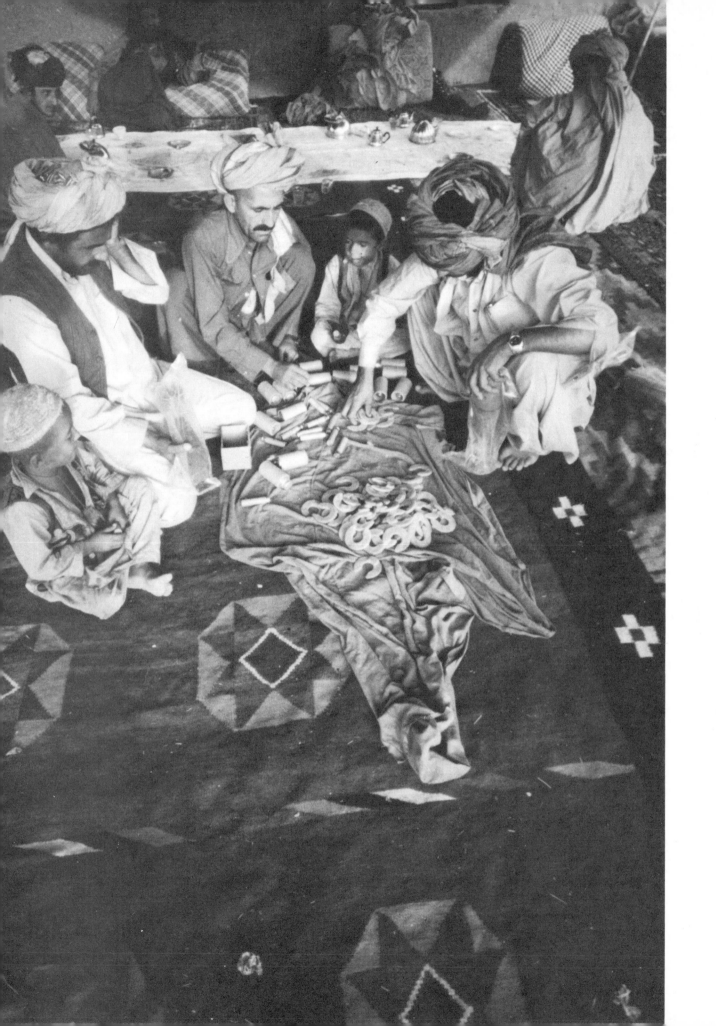

Above: *El Salvadoran rifleman returns fire during long fire fight at village of Amitlan Abajo. Troops commonly use forefinger as brace and index finger as trigger finger on an H&K G-3 rifle.*

Above, right: *Dr. Demo, staffer John Donovan, teaches airstrip security troops the use of Garrett electromagnetic metal detector.*

Update from El Salvador

This special report was filed from Central America in January 1984. It is one of the best overviews of the situation in El Salvador printed in America.

Between *Soldier of Fortune*'s first major trip to El Salvador in April '83 and the second trip in August '83, substantial changes took place, which we believe enhance prospects for El Salvador's future.

Any reasonably intelligent observer would conclude that the war being fought in El Salvador is not one for that country alone, but for the entire Central American region. Granted, there is internal dissatisfaction within El Salvador (and some valid reasons for it), but it is clear the war would not be raging at its current level without substantial backing by the Soviet Union (through its client states, Cuba and Nicaragua), giving the guerrillas money, supplies and advisers.

And it is equally clear that the guerrillas' long-range target is not El Salvador, but all of the Americas, including the United States.

Consequently, if we are going to ask Salvadoran grunts to fight this round for us, we should, at least, provide them with enough support to have a reasonable chance of winning—and a better than 50–50 chance of living if hit.

Because of the improved circumstances in El Salvador and the fact that SOF Publisher Robert K. Brown had identified additional areas in which SOF's A Team (eat your heart out, Mr. T.) could help out, he headed a 12-man SOF unit to El Salvador for three weeks in August.

SOF's full report follows.

SOF's Team

Robert K. Brown, Editor/Publisher of *Soldier of Fortune*, and lieutenant colonel, Special Forces, USAR. One tour in Vietnam, including command of the Tong Le Chon Special Forces Camp and experience as an Infantry Battalion S2.

Alexander M. S. McColl, Director of Special Projects, SOF. Colonel, Special Forces, USAR. Eleven years active duty, including two tours in Vietnam (SSO ACSI J2 MACV, District Senior Adviser, MACVSOG). Graduate of U.S. Army War College.

John Early, president, Albuquerque Parachute Center, Albuquerque, N.M., and a SOF Contributing Editor. Former captain, Selous Scouts, Rhodesian Army. Former captain, U.S. Army Special Forces. Four years, eight months and 13 days in Vietnam, including the siege of the Lang Vei Special Forces Camp. Expert in all aspects of parachuting and anti-terrorist operations.

Ben Jones, former major, Rhodesian African Rifles, former first lieutenant, USMC; former first lieutenant, U.S. Army; 24 combat jumps in Rhodesia; expert in anti-guerrilla and anti-terrorist operations.

Cliff Albright, retired Republic Airlines DC-9 captain (13,000 hours flying time, including about 400 in DC-3/C-47 type aircraft); commander, Phantom Division, Tennessee Airborne (a paramilitary/sport parachute organization); master parachute rigger, jumpmaster and instructor with 510 jumps.

John Donovan, owner, Donovan Dynamiting, Danvers, Ill., and SOF Contributing Editor, Major, Special Forces, USAR. Demolitions expert; after the U.S. government, his firm is the largest user of C4 in the United States. Twenty years of law-enforcement experience.

John Doe, weapons instructor, served with the USMC in Vietnam. First sergeant major in the Selous Scouts, Rhodesian Army. Noted weapons instructor and expert in anti-terrorist and long-range reconnaissance operations.

Peter G. Kokalis, SOF Military Small Arms Editor. Served with U.S. Army in technical intelligence branch. Distinguished writer, collector and expert on military automatic weapons.

Ralph G. Edens, president, Security and Research Ltd., Humble, Texas. International security consultant and unconventional-operations expert.

John Padgett, physician's assistant, certified. Three and one-half years as Special Forces medic in Vietnam, one year in Thailand. One year each in rural health clinics in Nicaragua, Micronesia and Alaska. Spanish linguist.

Philip Gonzales, family nurse practitioner. Two years as Special Forces medic in Vietnam. One year running health clinic in San Blas Islands, Republic of Panama. Spanish linguist and photographer.

Thomas D. Reisinger, assistant to SOF publisher. President, Refugee Relief International, Inc.; director, Parachute Medical Rescue Service (PMRS).

Summary of Activities

Early, Albright and Jones spent several days working with and improving the operation of the rigger loft of the FAS Airborne Battalion (*Battalion Aerotransportado*). Early and Albright had brought several thousand dollars' worth of parachute-related supplies, equipment and spare parts. As a result of their efforts, most of the airborne-related deficiencies and shortcomings noted in the last report were corrected, and training for riggers was initiated, including cargo-drop training. At the end of August when SOF left, the unit had 480 complete rigs ready to go.

Donovan conducted a follow-up demolitions course for the enlisted engineer sappers of the Atlacatl Battalion and an advanced demolitions course for three lieutenant instructors of the Atlacatl Battalion. He also inspected and inventoried the demolition materiel on hand at the Morazan Battalion at San Francisco Gotera.

John Doe conducted a series of three-day classes for FAS helicopter door-gunners and made arrangements to conduct future classes in basic combat-pistol shooting for FAS pilots, basic sniper techniques for the FAS Airbase Defense Battalion and advanced sniper-training for the Atlacatl Battalion. The door-gunner training resulted in significant improvements in the morale, motivation, weapons-malfunction rates and shooting skill of the unit.

Peter Kokalis completely overhauled the weapons inventory of the Atlacatl Battalion and continued his program of weapons instruction for that unit.

Early, Jones and Kokalis conducted a three-day ambush and counter-ambush training program for selected junior officers and NCOs of the Atlacatl Battalion.

A team consisting of Brown, Early, Jones, Edens, Padgett and Gonzales visited Lt. Col. Cruz, commander of Morazan Department, accompanying Cruz and elements of the Morazan and Airborne Battalions in the field to observe anti-terrorist operations. Medical activities by Padgett and Gonzales included saving the lives of two wounded airborne soldiers, Medcaps that treated several hundred ci-

Right: *Bob Poos, Executive Editor of SOF, gives victory salute after Salvadoran government troops drove off a large guerrilla force in an hour-long fire fight.*

Medevac chopper about to land in the village of Amitlan Abajo.

vilians and field-health and sanitation classes for Salvadoran troops and civilians.

A second team, consisting of McColl, Padgett, Gonzales and SOF Managing Editor Jim Graves, visited Lt. Col. Cruz and Dr. Alcides Caballero Lopez at San Francisco Gotera for further Medcaps and health classes for troops and civilians, which included assisting in the delivery of a child by Caesarean section, and evaluation of military training and civic-action projects.

Padgett and Gonzales also trained FAS helicopter door-gunners in basic life-saving procedures, which should help reduce the death rate of wounded troopers, conducted classes for enlisted medics of the FAS and the Atlacatl Battalion, and made a health and sanitation inspection of the Atlacatl base camp.

SOF Team Recommendations:

With minor exceptions, the recommendations in the April trip report still stand. SOF did note some improvement in the areas of medical support (chiefly the U.S. Army medical team under Col. Morales), civic action, psychological warfare and aggressive field operations in guerrilla-infested areas.

The root cause of most of the Salvadoran weaknesses is the miserly, restrictive support provided by the U.S. Congress to El Salvador.

The system of a standing MilGroup and MTTs that come in-country for short periods for specific tasks should be replaced by an integrated Military Assistance Command in which the working-level trainers are associated with their units on a long-term basis and the ban on U.S. trainers going into the field with their troops needs to be lifted, as well as the arbitrarily decided limit of 55 U.S. trainers in El Salvador.

The Airborne Battalion should be retrained and used as a national airborne fire force to react quickly to enemy attacks on isolated outposts, to gather intelligence information about enemy targets and to perform other tasks requiring more than a truckborne reaction force—which usually arrives too late or gets ambushed en route or both.

In any military operation and especially in a counter-guerrilla war, victory hinges on the ability to bring decisive offensive action against located and identified enemy units, base camps, supply installations and headquarters.

In anti-guerrilla warfare, the two most difficult aspects of this are timely and accurate target acquisition, and sufficiently rapid and flexible mobility for the attacking force so that it can attack and destroy the target before it gets away. The Salvadoran Armed Forces do have adequate target acquisition means. For movement of an anti-guerrilla strike force in this circumstance, the optimum

means are helicopters but in-country resources are totally inadequate to move more than about a platoon at one time, which is not a large enough force to do the job. Road-bound movement by truck is too slow and too exposed to ambush.

With not too major improvements in their aircraft maintenance situation, the FAS can muster enough air assets, chiefly C-47s and Aravas, to lift a company-sized force. Under these circumstances, airlift and parachute insertion are the only viable way to provide the required mobility to an anti-guerrilla strike force. This was the original rationale justifying creation of the airborne battalion. Unfortunately it has chiefly been used as a conventional infantry force and never fully trained and utilized in its intended role.

Left: *Preparing to move out in pursuit of communist guerrillas.*

Below: *El Salvadoran soldier from Alpha Co., 1st Bn., 5th Brigade, shows a rotted boot and exposed foot.*

SOF Trains Freedom Fighters in Nicaragua

This fascinating report, filed by James L. Pate in July 1985, follows the training mission of SOFers in Nicaragua.

We clear our last military checkpoint in Las Trojes, back on relatively safe ground, and head out of the little farm village back into the mountains. Just three days before, two U.S. State Department personnel headed for the FDN front in northern Nicaragua were turned back to Tegucigalpa at this roadblock. Access to the camps by journalists has also been limited.

But our SOF team—myself, Dye and Special Projects Director Alex McColl—have something the FDN badly wants, intelligence on field-expedient methods for taking out the Soviets' deadly Mi-24 helicopter gunship. That makes wafting us through a matter of priority for them.

"We are expecting one of two kinds of attack," Colonel Enrique Bermudez tells the SOF team during a briefing on our first morning in the FDN border command post. "Sooner there will be artillery attacks, or later on, if that doesn't happen, we'll very probably get hit by the Hind helicopters. That is where your people can be very helpful, by providing information on how to defeat this gunship. We also need training in how to set better booby-traps, mines and demolitions."

In his sessions with Bermudez and other FDN leaders, Dye emphasizes the need to fight unconventionally. When Bermudez expresses his need for helicopters, Dye acknowledges this requirement but reminds him that this never stopped the Vietnamese from transporting millions of tons of supplies down the Ho Chi Minh Trail. Perhaps it would be a good idea to make porters out of the men stuck in garrison because they lack weapons and ammo to actually fight.

Bermudez introduces us to Commandante Gustavo, CO of the FDN's Echo Company, officially designated *Commando de Operaciones Especiales*.

Below: With a folded copy of SOF's anti-helo ops publication under his arm, this FDN special Forces squad leader is absorbed in a reprint of the CIA guerrilla-warfare manual distributed through the good graces of Bob Brown.

Right: An FDN boot-camp bulletin board illustrated with cut-outs from the previous month's Soldier of Fortune. *Team member Dye takes note.*

Gustavo, a good-humored, thoughtful man, was a medical student before the revolution. He escorts us to his bivouac.

As Gustavo musters his men under the chow shelter, Dye unpacks the gear we brought to donate to the Special Forces company. This includes a Heckler & Koch flare launcher with long-range magnesium flares and a line-throwing gun manufactured by the Naval Company Inc. of Doylestown, Pa. While I pass out *Soldier of Fortune* reprints of the CIA guerrilla warfare manual, Dye circulates Spanish translations of the brochure he wrote on the Hind helicopter. It had been specially prepared for this mission by Art Director Craig Nunn.

Without being too specific, the brochure identifies the strengths and weaknesses of the Hind, as well as outlining its possible weapons configurations and providing pertinent technical data. The manual also contains well-defined ideas for using Hind weaknesses to destroy the aircraft.

The D-model provided to the Nicaraguans almost always mounts four pods, each containing up to 32 57mm rockets. It can also carry four laser-guided anti-tank missiles with ranges of up to 6.5 miles. Under the nose is a four-barreled 12.7mm machine gun. The Hind's high speed and the heavy, bathtub-shaped armor underneath make it almost impregnable to ground fire.

The Hind's three main weaknesses are its weak rotor head, its tendency to wallow in translational flight (moving from hover to forward flight or vice versa) and a hydraulic system that leaks flammable fluid profusely.

Freedom fighters in Afghanistan shoot down Hinds by positioning 12.7mm machine guns on mountain crests and firing *down* on the rotor heads as the helicopter cruises through valleys. There is very little armor anywhere on the top side of a Hind. Steel or Kevlar cable fired into the rotors as this bird moves into or out of an LZ can also bring an abrupt end to flight. That's what prompted employment of the line-throwing device.

Although the gunner and pilot ride in a pressurized cabin, the crew chief often opens the top half of the side hatch to provide increased ventilation and

Below: *New boots but no socks!*

visibility. That's a likely tactic in Nicaragua's muggy climate. A magnesium flare fired through this hatch may ignite leaking hydraulic fluid. An onboard fire means the pilot must set down immediately.

We spend all afternoon letting each trooper learn to load and fire the special weapons. Dye winds up by going over the Hind data once again and how it can be used against the *piricuacos*.

Above: *Commandante Gustavo fires H&K flare launcher as FDN Special Forces unit watch.*

Left: *Familiarization with a line-throwing gun donated by SOF.*

Dye breaks out his tinker-toy training aid the next morning to help instruct 12.7mm machine-gun crews in anti-aircraft gunnery. He's perversely proud of a wooden helicopter model he has doctored with tape to show the armor configuration on a Hind. We move over to anti-aircraft positions on an adjacent hill. Dye blanches over the position of these weapons—DShK-38s in sandbagged pits—and explains that they need to be redeployed from the topographic crest to the military crest of the hill and camouflaged. As the FDN has them deployed they make excellent targets for an air attack.

Field-expedient anti-aircraft sights are explained to the troops. We shape some with coat hangers and mount them on the guns. Using Dye's small helicopter, one trooper "flies" around the pit while others take turns learning to use the new sights that enable a gunner to properly lead his airborne target. It's not long before Echo Company has the hang of keeping the aircraft in the make-shift ring of the air sights.

Dye's class on the 81mm mortar is more abbreviated than AA training, but not by his choice. We fire illum rounds to conserve HE ammo and to avoid any possibility of firing into friendly patrols. A cease-fire is ordered when the FDN's G-2 notifies us that Sandinista patrols are in the area and there is a possibility they might use the flares to pinpoint our position. We use what little time is left showing the commandos how to dismantle Com-Bloc grenades and identify their fuse delay. We find a few zero-delay fuses and demonstrate various ways they can be used as booby traps.

Commandante Gustavo then proudly shows us mines his men have recovered with the help of a Sandinista deserter. The anti-personnel devices, which had been planted to kill freedom fighters, will now be turned on the communists. Cyrillic markings indicate the plastic explosive and detonators were made in Russia, while the wooden containers were of Cuban origin.

We present eager Echo Company members with SOF patches. As a token of his gratitude, Gustavo gives Dye a propaganda biography of Augusto César Sandino, dead guerrilla and figurehead of the communist revolution in Nicaragua. Gustavo, who inscribes it to Dye, had taken it off the body of a *piricuaco* which he had dispatched to the realm of all good communists.

Left: *Colonel Enrique Bermudez, field commander for the Democratic Force of Nicaragua.*

Above: *Dye uses training aid to teach FDN's 12.7mm AA gunners.*

Left: *Guatemalan paratrooper after a week in the mountains.*

Guatemala: Political Perspective

In March, 1983, Robert J. Caldwell summarized the political situation in Guatemala.

Guatemala has been run by generals ever since a CIA-backed coup in 1954 overthrew the left-leaning government of Jacobo Arbenz. Guatemala is still run by a general—Jose Efrain Rios Montt. But Rios Montt is a general with a difference, several in fact.

In a predominantly Roman Catholic country, Rios Montt is a devout, evangelical Christian. In a country where most generals run the political gamut from deeply conservative to overtly reactionary, Rios Montt is something of a moderate. Above all, Rios Montt seems to recognize that the communist-led insurgency plaguing his country is not solely the product of Cuban-inspired subversion, although many of the guerrilla *comandantes* have in fact received training in Cuba.

Rios Montt retired from the army and ran for president in 1974 as a candidate of Guatemala's Christian Democratic party. He lost, accused the government-backed candidate of stealing the election, and was promptly packed off to de facto exile as military attache at the Guatemalan Embassy in Madrid.

Shortly after he returned to Guatemala in 1978, he renounced his Roman Catholic faith and joined the evangelical Church of the Word. The church, known as *Verbo* in Guatemala, is a branch of the California-based Gospel Outreach.

By early 1982, Rios Montt was devoting full time to his church, working as administrator of *Verbo*'s primary school in Guatemala City. And he wasn't pleased with what was happening in his country.

Marxist insurgents were gaining converts among the impoverished Indians of Guatemala's highlands. The government was responding with growing repression.

Terrorism from the left and right was reaching epidemic proportions. The Carter administration, reacting to widespread reports of human-rights violations by the government, suspended all military aid and sales to Guatemala.

When the government-supported candidate was declared the winner of presidential elections last March, Guatemala seemed headed for four more years of lawlessness and escalating insurgency. It was then that a group of young army officers, convinced that government corruption was losing the war against the *subversivos*, organized a coup.

Troops backed by tanks and artillery surrounded the national palace on the morning of 23 March.

The lame-duck president, Fernando Romeo Lucas Garcia, agreed to surrender but only to a senior officer.

The coup makers telephoned Rios Montt. He enjoyed a reputation for incorruptibility and had earned the respect of many junior officers during his tenure as commandant of Guatemala's military academy in the early 1970s.

Rios Montt agreed to negotiate the government's surrender and was in turn appointed chief of the new junta. He offered amnesty to any guerrillas who laid down their arms and surrendered. He promised unrelenting war against those who did not. He also promised a crackdown against left-wing terrorism and right-wing death squads, and an end to government abuses of basic human rights. He dubbed his strategy for prosecuting the war *"fusiles y frijoles"*—"bullets and beans" in English. It combined a renewed military offensive against the guerrillas with civic action programs designed to regain the allegiance of disaffected Indian *campesinos.*

Right: *Like this NCO, 80 percent of all Guatemalan soldiers are Indian rather than* **Ladino.**

Above: *Dominican governments' entire air force, obsolete U.S. P-51s, line up at San Isidro Air Base. On May 13, the rebels lost their most valuable propaganda machine when these planes knocked out the radio-TV station.*

Lt. David Humble, 82nd Airborne Division was XO for an artillery battery that was sent in and never allowed to fire a shot. Total U.S. losses as of May 20 stood at 20 KIA, 102 WIA, and one missing.

Santo Domingo: A Domino That Didn't Fall

April 24, 1965, Santo Domingo explodes in a communist-inspired rebellion. One more domino teeters on the brink of collapse. President Johnson announces on television that the United States *will* intervene to prevent another communist takeover. There will not be a "second Cuba."

The 82nd Division is put on full alert. The U.S.S. Boxer and a contingent of 1,100 leathernecks close in, while 40 other vessels isolate the tiny Caribbean island.

"The American nations cannot, must not and will not," Johnson was to say, "permit the establishment of another communist government in the Western Hemisphere."

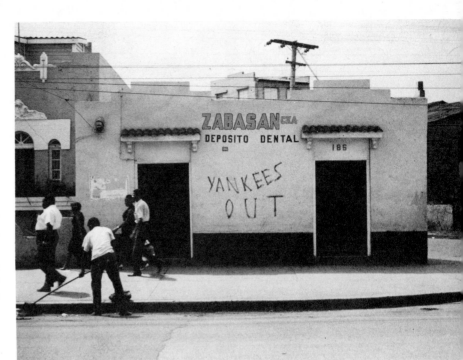

Above: *The two shots above show how the American intervention was both welcomed and rejected. Civilian casualties numbered into the thousands.*

Above: *After the fighting cooled down, U.S. trucks delivered rice, beans and powdered milk to starving civilians. Children fought each other over C-ration scraps.*

Below: *By the end of the first week some 10,000 armed rebels had been surrounded by 24,000 American Marines and Paratroopers. Some of the house to house fighting was intense, and at one point hundreds of bodies littered the streets. But for most of the GIs who were dug in, the situation was only made worse by the monsoon rains.*

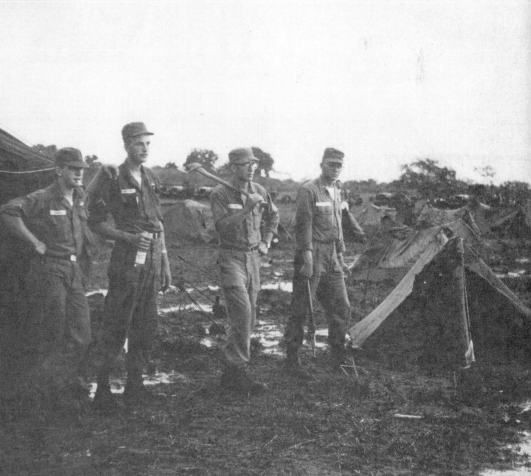

SOF in Grenada

First published in SOF in early 1984, Jim Graves reports on the intelligence gathering mission of the SOF staff immediately after the Grenada liberation.

"Where's the war?" the journalists demanded as they rushed through the streets of Grenada's capital, St. George's.

Smiling Grenadians answered in their singsong English, "The Cubans have gone to the hills. Welcome to Grenada. Are the Americans going to stay? We want them to. It's a good thing you didn't wait a few more days."

By Sunday, 30 October, the liberation of Grenada was almost complete.

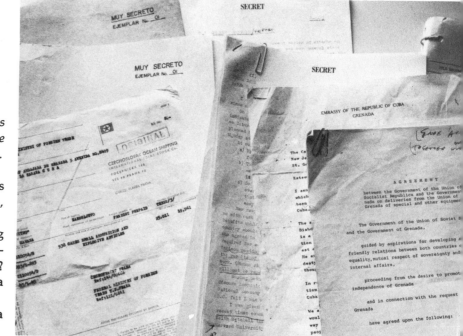

Above: *Some of the documents SOF scored on its intelligence gathering mission.*

101

Left: *Capt. Blatti, Operations Officer, 1st. Bn., 505th Infantry, 82nd Airborne, coordinates "Blackhawk" flights to LZs in search operation for Cubans on Grenada.*

Below: *Czech M24 SMG, Federal 203A riot gun, M3A1, Sten and Sterling SMGs, and an AK-47 with recent-issue composition magazines.*

It was a liberation, not an invasion. Or at least that's the way the Grenadians saw it when *Soldier of Fortune*'s team (Editor/Publisher Robert K. Brown, former Ranger Rod Hafemeister and Jim Graves) arrived on Grenada.

Although rumor had it that significant numbers of Cubans were retreating into the hills to conduct guerrilla operations, in the next 48 hours, SOF's team observed only six recently captured prisoners and only three of those were Cubans.

En route to St. George's from Point Salines Airport, through swirling dust kicked up by aircraft (C-141s, C-130s and helicopters) swarming around the strip and motor vehicles (either hauled in from the United States or confiscated from the Cubans and relettered "USA" by America's 6,000-man landing force), we observed few signs that a military action had taken, or was taking, place.

In a quick tour of the area around Point Salines airport, we observed an artillery battery from the 82nd Airborne on the northeast end of the runway,

Left, above: *R.K. Brown checks the in-basket on the desk of Grenada's Deputy Minister of Defense.*

Left: *Field gear storage room with new Soviet canteens, packs, mess kits and entrenching tools.*

some troops on the perimeter, some foot patrols along the road, troops in vehicles on the road, numerous vehicle checkpoints manned by 82nd Airborne troops, some abandoned Cuban anti-aircraft guns and several shot up BTR-60s, but it was obvious that any real resistance had long since crumbled.

SOF's team got to the island with the first big load of press (there were about 160 reporters) five days after D-day. Once the mad rush for the limited number of rooms in the St. James Hotel was over, the press hit the streets to find the war and Grenadians to interview.

Most of the press was infuriated about being kept off the island during the fighting and quite a number of them wanted to prove America wrong in taking the island—exclude SOF from that category—so the next day or two came as a shock to many in the press corps. Grenadians everywhere on the island greeted reporters with the same smiles and question: "Are Americans going to stay? We want them to."

With the invasion action virtually completed, the only thing for SOF to do was to piece together from the troops what had taken place.

A quick trip to Ft. Rupert, just about 100 yards up the hill behind the St. James Hotel, by *Washington Times* reporter Jay Mallin, Lionel "Chu Chu" Pinn, an old friend of SOF, and myself uncovered the fact that no one was guarding the NJM Central Committee headquarters, the deputy minister of defense's office nor the equipment stores at Ft. Rupert.

On Monday morning, Mallin and the SOF team searched all three locations. In addition to a fine collection of new Soviet helmets, canteens, mess kits, packs, AK-47 bayonets, military manuals and

Above: *Troopers from B Co., 307th Infantry, 82nd Airborne display guns found in five warehouses near Grenada's airport.*

Right: *U.S. OH-60 "Blackhawks."*

Below right, facing page: *Soldiers take shelter at a Cuban base used for training Pan-American guerrillas.*

Below: *Soviet BTR-60PKB where the 75th Rangers left it.*

the NJM flag that had flown over the fort, SOF picked through the papers scattered around the office of Lt. Col. Ewart Layne, Grenada's deputy minister of defense.

The intelligence finds (we also located documents in Ft. Frederick and Butler House, the prime minister's office) were significant: shipping manifests of weapons from the USSR, through Cuba to Grenada; a defense treaty between the USSR and Grenada; a roster of Grenada's militia; a summary of Political Bureau meetings; a top-secret report from a Grenadian double agent who was trying to infiltrate a CIA operation on Barbados; a counter-intelligence summary; rosters and correspondence relating to the training of Grenadian PRA members in the USSR, Cuba and Vietnam; a training agreement between Grenada and Nicaragua; and miscellaneous correspondence from Cuba, including one letter from Fidel Castro to the Central Committee written after Bishop was arrested.

In general, the documents and other physical evidence observed by SOF and information gleaned from other documents recovered by the U.S. forces and released in November indicated that: 1) Cuba and the USSR were turning Grenada into a strategic military base; 2) as in Libya and Nicaragua, more weapons than Grenada could ever use had been shipped to the island; 3) Bishop was killed because of a power grab by Coard and because he was not as pro-Cuban as other Central Committee members thought he should have been; 4) the NJM was losing control of the country because of its excessive pro-Cuban and pro-communist attitude; and 5) some well-known Americans had highly questionable dealings with the NJM.

Once SOF gathered its intelligence reports from Grenada, we still had deadlines to meet. After the mad rush to get our stories, most of us journalists faced another scramble for transport back to the United States to get photos processed and stories into print. Back in the States, after being cut off from news for almost a week, we learned of the international and national reaction to the Grenada rescue mission.

On the international scene, the United Nations, including those member states who are our allies, "deplored" the American action in Grenada.

Meanwhile at home, during the operation, in Miami the Grenada National Steel Band entertained a crowd of Grenadians, who were celebrating the liberation of the island. Two days later, a crowd of 3,000 Latins turned out for a flag-waving rally where the most popular signs were "Viva Reagan" and "First Grenada, Nicaragua Later and Cuba Third."

According to a nationwide poll, the American public enthusiastically favored the operation—by a 9-to-1 majority—primarily because the American students evacuated off the island returned with thanks for their "rescuers."

Relieved medical student Jeff Geller summed up the students' feelings in thanking Reagan and the American troops who were on Grenada: "Prior to this experience, I had held liberal political views, which were not always sympathetic with the position of the American military. There's one thing to view an American military operation from afar and quite another to be rescued by one."

Even some Americans who could not be expected to concede that the operation was the wise thing to do did so. Democratic Congressman and Speaker of the House Thomas "Tip" O'Neil, who had been an outspoken critic of the operation, said he had changed his mind after hearing the report of the

congressional delegation that went to Grenada. SOF suspects he actually saw the poll results indicating the overwhelming popular approval of the action.

CBS News announced it had polled 304 Grenadians and found that 91 percent "are glad the U.S. troops came to Grenada," while only eight percent were opposed. Also 76 percent of those polled said they believed Cuba wanted to take control of Grenada's government, and 65 percent said they believed the airport was built for Cuban and Soviet military purposes.

On the liberal front, columnists Anthony Lewis, *New York Times*, and Mary McGory, *Washington Post*, questioned the Grenada operation. Of course.

Congressman Dellums, a member of the congressional delegation that went to Grenada, stated that he disagreed with the majority conclusion that the venture was justified as a means of protecting American lives.

SOF strongly suspects that Lewis, McGory and Dellums are not happy with the liberation of Grenada. But the Grenadians, the American public and SOF are.

Right: *The slogan is about all that remains after an AC-130 blasted this building.*

Left: *Cuban hero waits for transportation to P.O.W. pen.*

Above: *Soldiers of the 307th Infantry, 82nd Airborne display their color and a captured People's Revolutionary militia flag.*

Dominoes in Peru

These insights into guerrilla activities in Incaland were reported in SOF in March, 1983, by Jay Mallin.

Above: *Tracts and dynamite are the weapons of "The Shining Path."*

Near midnight on 3 March 1982, heavy gunfire erupted in Ayacucho. A force of 50 to 60 guerrillas attacked the CRAS (*Centro de Rehabilitacion y Asistencia Social*). The CRAS is a prison at the northern end of the city, a brownstone, almost windowless structure, about a block long with but one entrance. Firing weapons and hurling sticks of dynamite, the guerrillas blasted their way inside. The firing lasted about an hour. When it was over, some 257 prisoners, including 10 women, had been freed, among them a number of suspected guerrillas. Ten men were killed, including two policemen. The other casualties were guerrillas and prisoners. The interior minister, retired Air Force Gen. Jose Gagliardi, reported: "The assault was carried out by well-prepared people and was similar to a military commando attack."

No one knows how many *Sendero* there are. The vice-minister of interior says, "They are not numerous." He places the membership at between 2,000 and 3,000, "all very active." These figures are higher than the usual estimates of *Sendero* strength; the minister may be including supporters of the movement as well as actual guerrillas. Probably the best estimate of active guerrillas, primarily in the Ayacucho area, is between 400 and 600 men. *Guardia Civil* Gen. Hector Rivera Hurtado, commander of the combined police forces at Ayacucho, states that guerrillas in his zone of operations number 150 to 200. He says that they operate in small groups but bring sympathizers along when staging attacks to create an impression of greater numerical strength.

The attack on the prison was the climax of a terror and guerrilla-warfare campaign that has been increasing in intensity since 1979. The explosion of bombs, theft of weapons, blowing up of power lines, attacks on rural police posts, killing of specific individuals (landowners and suspected police informers) have all been part of this campaign. These actions have not been limited to Ayacucho but have occurred in Lima and other parts of the country as well. Indications are that the rebels are beginning to intensify their activities throughout Peru.

The reasons for the guerrilla movement in the Ayacucho area seem clear: economic stagnation combined with the university serving as a front for Marxist thinking and action. But are these the only reasons? The question asked both by Peruvians and

Facing page: *Former Senderista killed by a **Guardia** sweep through an Ayzarca ranch.*

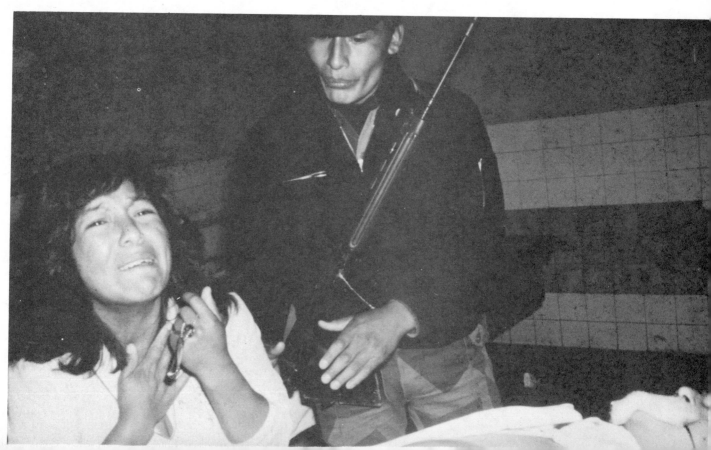

Below, left: *Wife of a Civil Guard and her wounded husband. The toll can be heavy.*

Below: *One the law caught up with.*

knowledgeable foreigners is: "Why Ayacucho?" Despite the *Sendero*'s apparent lack of connection with any foreign country, was Ayacucho actually picked—perhaps by Havana—because of its symbolism, since it is the place in which Simon Bolivar defeated the Spaniards in 1824, ending Spanish rule in South America?

Is this the beginning of an effort to turn into reality Fidel Castro's prediction years ago that the Andes could be the Sierra Maestra of South America? At least one ambassador in Lima sees this as a "sinister" possibility. Peru has a particularly strategic position on the continent: It borders on five countries and could serve as a breeding ground for continental subversion, much like the subversive/guerrilla-warfare center into which Guevara hoped to convert Bolivia.

Cuba has given guerrilla training to Peruvians. To date, however, no proof has emerged of a Cuba-*Sendero* link. At present such an association may not be ideologically possible. But as the *Sendero* grow they may well reach a point—say if the Peruvian Army enters the fray—where they will need foreign assistance. And what more likely place to look for it than Havana?

Thailand's Battle Road

SOF Staffer, Jim Coyne, filed this on-the-spot report in February, 1982.

The first thing Col. Surachet did after we landed, the very *first* thing, was issue me an M16 and a tiger-striped vest with ten 20-round magazines. "You'd better take it," he said almost apologetically, "the area isn't 100-percent safe."

He smiled as I took the weapon. I thought: "He ought to know." The M16's bluing was hand-worn to a dull steel finish from years of use and cleaning.

I hadn't carried an M16 in a combat area since Vietnam, but it felt as natural in my hands as the cameras I have carried since.

In the distance, the sharp *crump* of an exploding grenade echoed and rolled through the green mist-shrouded hills from the direction of "the road." We walked to a sandbag-reinforced bunker dug deep into the ground and bristling with radio antennas for a brief situation report.

In April 1975, after the collapse of Cambodia and South Vietnam and the capitulation of Laos to the communists, the situation in Thailand—and all Southeast Asia—turned critical. The first dominoes had fallen. Captured plans disclosed that the strategic goal of the CPT (Communist Party of Thailand) was to seize a vast, rich, L-shaped section of northcentral and northeast Thailand extending south from the Laotian border, through the central highlands, then eastward to Cambodia.

Only effective and immediate action could prevent this from happening. Military operations alone would be inadequate. Long-range planning was required to solve the broad social, political and economic problems facing Thailand. King Bhumibol Adulyadej of Thailand outlined a comprehensive Strategic Development Project in 1976.

Known privately as "The King's Project," it used—and continues to use—the combined talents of Thai business and military leaders, educators, farmers and students to deny areas of influence to the CPT. The project established aggressive rural programs of land reform and agricultural guidance in disputed areas.

Within the sanctuary of triple-canopy jungle and difficult terrain, the CPT operated with virtual impunity, subjugating villages, taxing the population and directing terrorist attacks against the Thai government. The CPT infrastructure included a politburo, weapons and ammunition-storage depots, repair and maintenance facilities, radio transmitters, cadre training areas and a large, well-equipped hospital. The heavily-armed mainforce units (515 Company, 520 Company and 523 Company, 130 men each) exploited the population of 10,000 Meos (H'mong) and Thais with Marxist discipline. If the CPT continued to consolidate the northcentral highlands, the fertile cultivated provinces to the south and then Bangkok itself would be threatened.

The first large-scale military operation against the CPT stronghold and northern headquarters began in early 1977 with operation "Pa Muang Padetsuk I." It encountered stiff resistance. The CPT fought from long-held fortified positions in areas favorable to defense. In the beginning, only moderate gains were accomplished. The emphasis, however, was on long-term strategic control and development, not search-and-destroy tactics. Gradually, the process of consolidation began to turn in the government's favor.

In 1979, the decision was made to strike straight at the heart of the CPT redoubt: the cultivated valleys and growing areas which provided food for the communists near Nong Mae Na, and the strategic heights of Khoa Ya and Khoa Kor. A four-phase plan sought to deny, once and for all, the "ocean" in which the CPT "swam."

Intelligence-gathering operations, combined with limited but effective small-unit military actions, ended just prior to an assault on the final objectives. Weeks before the first CA (combat assault), aircraft, helicopters, troops and support materiel began to arrive with great secrecy, at the small Royal Thai Army airstrip at Pitsanuloke. The staging area was cordoned off, and the troops quarantined. Aircraft arrivals, staggered at long intervals during daylight hours, created an illusion of normalcy.

In the predawn darkness of 24 January 1980, aircraft engines and helicopter turbines suddenly shattered the silence of the small airstrip. Airborne and helicopter assault troops harnessed up and began last-minute equipment checks prior to take-off. The first CA of Operation "Pa Muang Padetsuk II" committed four battalions of the Third Army Region (including airborne, ranger, special-forces units, infantry and armor) in support of a large road-construction project which would eventually encircle the entire tactical Area Of Operations.

With rented and leased D-9 and D-6 bulldozers, double-8 × 8-lumber bolted to the dirver's cage for protection from small-arms fire, and supported by APCs and infantry, they did it: They built the road.

For three months the army punched simultaneously against all known targets, in two directions, through walls of virgin teak and bamboo which had once been an impenetrable National Preserve. The CPT pulled out all the stops.

As the bulldozers continued to cut deep into CPT territory, the communist terrorists began to slug it out toe to toe in a series of vain, costly attempts to

Right: *Pushing through on patrol.*

Above: *Thai trooper moves through jungle during Khao Khor ops.*

Left: *Searching a CT hut with machete and M16.*

Above: *One way to get around.*

stem the advance. In a captured document taken from a command bunker near Khoa Kor, the CPT admitted that if the strategic areas of Nong Mae Na and Khoa Kor fell, the highlands would be lost. In desperation, they fired 57mm recoilless rifles from the shoulder, mined the roads (60 percent of Thai casualties came from mines) and hurled themselves against the steel blades and bullets of the Thai army. They lost.

The lead clearing elements of the road-construction teams linked up successfully in March 1980, completing the first phase. A road now encircled the area of operations. It could not remain a dirt road; soon the monsoons would arrive.

Far below, under the glare of lights, road construction continued. "We're building more than a road here," Gen. Pichitr had said to me earlier. "We're building the future of Thailand."

Philippines: War on Two Fronts

Robert J. Caldwell reported on the "shadow war" in the Philippines in September, 1982.

At first glance, the Philippines doesn't look like a country at war. Manila moves to the rhythm of commerce: commuters and throngs of visitors from abroad. No soldiers on the streets, no military parades, no loudspeakers, no patriotic exhortations.

The second look is more revealing. In many of the capital's larger office buildings, private security guards screen all packages and issue on-premises passes in exchange for a driver's license or other identity document. At some of the five-star tourist hotels, visitors may be asked to pass through metal detectors and to submit to a peek inside their briefcases or purses.

Right: *Bob Caldwell with "Centurion" strike force about to go on a sweep in search of MNLF guerrillas.*

Above: *Strike force commander Lt. Alindayu maintaining contact with battalion headquarters.*

The domestic terminal at Manila International Airport is guarded by police carrying pump-action Remington shotguns. And Philippine Airline planes on the tarmac are watched by uniformed members of the paramilitary Philippine Constabulary who tote M16s.

Still, the casual tourist might assume that this is nothing more than sensible precaution against the sort of terrorism that can happen almost anywhere. And, in part, they would be correct.

But the Armed Forces of the Philippines (AFP) are at war. Indeed, they are fighting two wars—one against communist guerrillas of the New People's Army, and the other against Moslem insurgents under the banner of the Moro National Liberation Front. The NPA, the military arm of the Communist Party of the Philippines, gets no known outside support. The MNLF, however, has received both arms and some political support from Moslem nations, notably Mommar Khadafy's Libya.

Both of these insurgencies are largely invisible to the tourists and even to many Filipinos living outside guerrilla-infested areas. The small-scale encounters that characterize both wars typically occur in rural regions, often in the less-developed and more remote sections of the Philippines.

Left and above: *Equipment shortages are common even in this elite company of the 5th Battalion; most troopers wear plain fatigues rather than cammies.*

Casualties in any given action are rarely enough to rate more than a few paragraphs in Manila's daily newspapers, or the barest mention on radio or television newscasts. And yet, the 10-year-old Moslem rebellion has claimed an estimated 50,000 lives. As for the stubborn insurgency waged since the late 1960s by the New People's Army, the Philippines is the only pro-Western country in Southeast Asia plagued by a communist guerrilla movement that is actually growing in strength.

NPA forces, estimated at up to 6,000 armed, full-time guerrillas, are concentrated in northern and southeastern Luzon, on Samar in the Visayan Islands and, increasingly, on Mindanao.

The estimated 10,000 armed guerrillas of the se-cessionist Moro National Liberation Front operate on Mindanao, Palawan and the Sulu Archipelago.

In response to these threats, the Armed Forces of the Philippines have doubled in size since 1972. At present, the AFP numbers about 112,000 regulars, 43,000 paramilitary Philippine Constabulary and 65,000 civil home-defense guards.

Getting a first-hand look at the AFP's counterinsurgency operations can take some doing. Understandably, the government of Philippine President Ferdinand E. Marcos is not eager to promote the kind of foreign press coverage that would discouarge tourism, foreign investment or international confidence in the country's stability.

Left and above: *Militiamen and their village.*

Bloody Belfast

This report was filed by David Mills and Rick Venable in July, 1984.

At HQNI (Headquarters Northern Ireland), a press spokesman was at first not happy to see us. A little discussion revealed that numerous British newsmen had been visiting Northern Ireland and making it sound like a cross between Tet and the Normandy invasion. When it comes to sensationalism, Brit journalists seem to be even worse than Americans, which I had thought impossible.

The view of the British government, which happens to be perfectly correct, is that nothing very sensational is happening here. Life in most of the city is poor, cold, wet, grim, unpleasant—but all of these are, for Northern Ireland, normal. Except for an occasional patrol driving through, I saw little to indicate that anything more serious than industrial decay was taking place.

Most of the trouble is concentrated in a few small areas, and even there not too much happens—usually. It is only the professional watchfulness of the Brits that keeps things calm. The IRA hasn't repented. It just can't get away with much these days—unless the Brits let their guard down.

We ducked quickly through the steel doors of a Brit position downtown. If you stand too long in the wrong place something, probably in .308, may happen. The bad guys know the Brits have to go through these doors and try to put snipers in positions to cover them.

Nine o'clock on a cold, dark morning. The outfit we were visiting was the Royal Regiment of Wales. Several men in combat gear were milling around the concrete courtyard, kidding with each other. I noticed this cheerfulness again and again: After months cooped up in fortresses, in god-awful weather, they remained alert and good-humored.

Armored vehicles stood dripping in the murk. Most were Humber one-ton trucks upgraded with armor. Rifles were the L1A1 SLR (self-loading rifle) with two-power scopes. When we got out on patrol, I realized that for picking a sniper out of a window at 300 meters, a long gun with glass sights rather than one of today's squatty assault rifles was the right way to go.

We went into the duty room with our escort officer, a captain. The walls were covered with maps, and radios stood on tables. On the walls were several television monitors attached to cameras that could be pointed and zoomed to show what was going on outside. I won't detail what the Brits have in the way of surveillance and control, since I don't really know what information the bad guys might find useful.

A blond sergeant showed us the area we would patrol. He knew every inch of that map by heart, what kind of neighborhood each was, what had happened there recently, what the buildings were. These guys definitely know their business.

Another trooper explained, "The worst neighborhoods are the poorest ones. As soon as a neighborhood gets rebuilt and the people have decent houses, trouble seems to diminish. I think a lot of the problem here is that the economy is so bad."

After a cup of coffee we went down to the vehicles where the patrol was waiting. We climbed in, camera bags catching on everything. The heavy-steel rear doors clanged shut. The cab is armored with swingdown shutters. The two men in the rear of the troop compartment poked their rifles out the rear slits. I noticed they kept their fingers on their triggers and a careful eye out the back. Unlike American forces, they don't fool around.

The engines cranked up, and we pulled into traffic. Having been on patrol with the Marines in Beirut some months earlier. I found myself carefully watching cars that pulled up behind us. I saw two important differences: While the Marines patrolled in open jeeps, the Brits stay behind armor—and cars behind us were looking into the barrels of two rifles.

When the Brits took to the streets in Northern Ireland in 1969, their street-patrol and riot-control tactics were more appropriate for a Gilbert and Sullivan opera than modern civil disorder. Platoon-sized squads patrolled in a box formation which protected a justice of the peace, a photographer, a reporter and a pair of banner carriers. When the squad met a mob, the unit commander hailed them through a loudspeaker, ordering them to disperse. The banner carriers held up their placard so that the side reading, "Disperse or we will use gas," faced the crowd. Then the justice read the riot act, and the photographer and reporter began to record the developing incident.

Left: *Delivering mail the Ulster way.*

Below: *The infamous "baton gun," Enfield's LG 7A1 riot gun saves lives by supplementing conventional small arms fire in riots.*

Left: *Royal Welsh trooper passes some free advertising.*

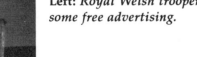

Squad troops were equipped with riot shields, batons and standard weapons (SLRs). If the rioters were merely throwing garbage or stones, the British were too gentlemanly to fire warning shots. Only if the crowd grew uglier the commander would order the front rank of his squad to take up firing positions. First, however, he pointed out the ringleaders and had the banner men reverse their placard: "Disperse or we will fire."

If this stunning display of force didn't work, the commander identified a specific target and ordered the men of the front rank to each fire a single shot. And it was truly a single shot—all weapons but one were loaded with blanks. (This practice was supposed to prevent identification of the shooter.) The file then ejected the spent cases. The theory held that the crowd would then disperse.

Naturally, these patrols became popular targets for snipers who took cover behind the crowd and vanished after firing at the squad. But the British soon became streetwise and developed the tactic of "satellite" patrols, as I discovered when we reached the patrol area.

We piled out of the vehicles—20 soldiers and one lone cop from the RUC (Royal Ulster Constabulary). The cop was important. Because the British were uncomfortable with military rule in Northern Ireland, they restored civilian government as soon as possible. Therefore the cop was at least formally in charge here. The soldiers were to protect him as he went about his ordinary duties: handing out parking tickets, rescuing cats, delivering summonses. And that's what he did. It takes guts to be a cop here. Policemen don't live in fortified positions. And they are vulnerable to attack at home.

The way a patrol works is this: The 20 troops divide into five "bricks" of four men each, all linked by radio. The command or primary brick walks the cop's beat with him, keeping him in the middle, while the other four bricks form a multi-brick which patrols a block away. These satellite patrols protect the central one and, in practice, a sniper can't get off more than a couple of shots before the patrols have sealed off the area. The multi-brick also protects each unit from snipers when crossing intersections. In south Armagh the troops patrol on the double along preplanned but apparently random routes in which they have worked out exactly where each man will take up his firing position to protect the others when crossing open areas and intersections.

Bricks are usually commanded by corporals. In fact, Northern Ireland has been called the corporal's war. The CO can also call up some "piglets"—armored Land Rovers—to surround his brick. "A ring of steel," as one officer called it.

The neighborhood we were patrolling was grim—blocks of flats with windows, warehouses with windows, high-rises with windows, commercial buildings with windows. You notice which windows are open. People wandered by, barely looking at us. We looked at them—carefully.

Fortunately the IRA is more vicious than courageous. An Iranian nutball would walk up and blow a trooper away, but the IRA is afraid of getting caught. They tend to go for soft targets—off-duty cops at home, for example. The doorbell rings, the guy opens it, and blam, he's dying in front of the TV with his kids watching. If you have any romantic ideas about the IRA, forget them.

The day was drizzly, the light bad. The bricks fanned out to take their positions, each man staying well apart from his buddies. They wore flak-jackets. I took out my bright-blue, very unmilitary shirt tail and tried to look conspicuously like a gringo photographer. I wish these guys well, but it's their war.

They fight it well. The men never bunched up. They stayed alert, squatted behind cover whenever it was possible, and kept an eye on the windows. This is a mark of good troops. Staying careful is easy when you are getting shot at a lot. The hard thing is staying careful week after week, month after month, when nothing much happens. You get a little bored, a little slack. Then comes that hypersonic crack, and somebody bleeds to death on the way to the hospital.

We came to a warehouse with several rough-looking characters standing around. They didn't like it when our patrol approached them to check them out, nor did they like our cameras. Well, nobody promised them a rose garden. The troops were courteous enough, and I suspect they would have been even without press around.

The real danger here is bombs. The bad guys will plant a command-detonated bomb and wait for a patrol to come by. The Brits do enough patrolling in the same areas that sooner or later they are likely to go by the wrong place. Boom! Further, the bomb is likely to be detonated from out of sight. For example, the switch might be in an attic somewhere, while the bomb itself might be two blocks away. The bomber watches a man on a corner in sight of the patrol. When a Brit passes the bomb, the man on the corner scratches his chin, which signals the sapper to close the switch. If he gets a few kids, well, the IRA figures they are expendable.

Perhaps the best protection against this is the checkpoints around the city that cordon one area from another. This keeps the terrorists from readily moving explosives any distance. It works pretty well, but not perfectly.

What every grocer needs.

During our stay, various Brits made observations I found interesting. An officer told me that they find line outfits, such as the Royal Regiment of Wales, to be better than elite groups for duty in Northern Ireland. The elites are aggressive by training and character. When a Brit gets killed, he told me, "Their instinct is to go out and waste somebody." They are too disciplined as a rule to do it, but they are more likely than leg outfits to interpret things in favor of shooting. The last thing the Brits need is to kill an innocent and stir up the population.

And they say that the IRA is more professional than it used to be. There are fewer of them, and they make fewer attacks, but the ones they make are more organized and the technique more sophisticated.

The rain grew heavier. People still walked the sidewalks, leaning into the drizzle. It was not quite noon, but by the light it seemed more like late afternoon. A Brit said hello to a couple of small kids we passed. They didn't answer, either from shyness or hostility. In parts of Belfast people are friendly to the Brits at night, when the IRA can't see them. By day they keep their distance.

The cop walked onto a porch and rang the door bell to deliver a summons. A soldier crouched at the bottom of the steps and looked carefully down the street. Water dripped from his chin. My fingers were numb with cold. A crazy sight: the cop, utterly British-looking and civilian, with a countersniper rifleman behind him like something out of I Corps.

I walked off into the rain. Nothing much had happened. Nothing much usually does. For me, there would be one more patrol in the afternoon, and I'd be on the air shuttle to Heathrow. These guys had lots more patrols in front of them. Days and weeks and months of patrols, inevitable losses, and little to show for it. They're up to it.

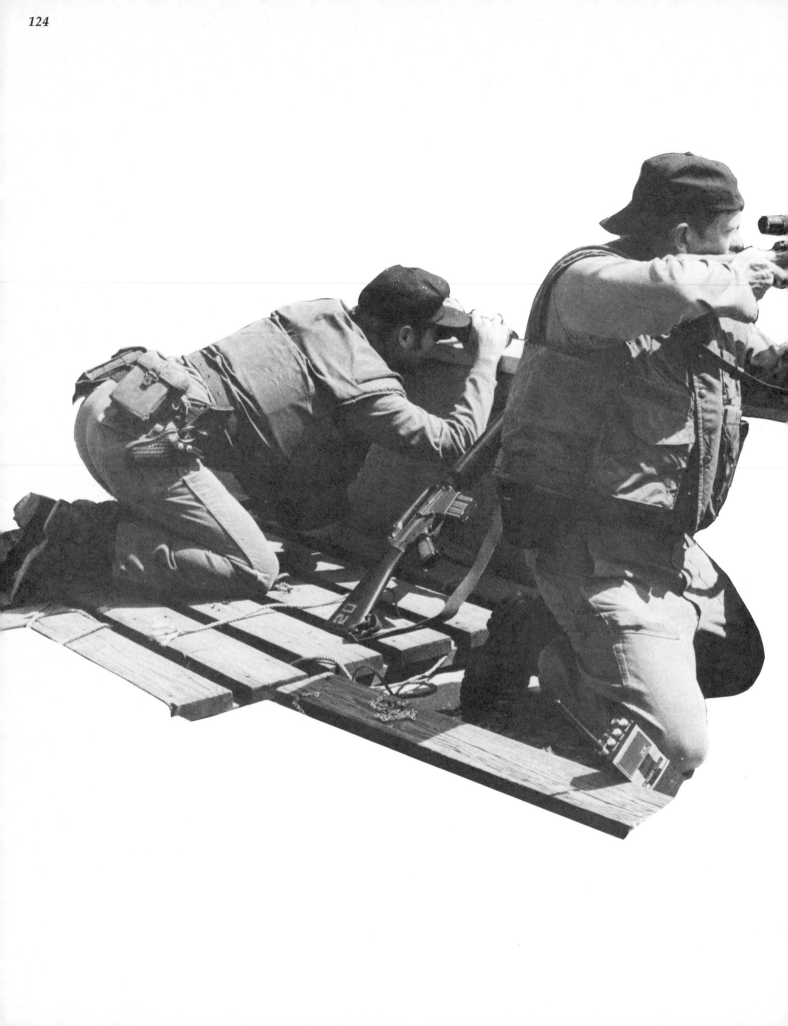

Weapons, Tactics and Equipment

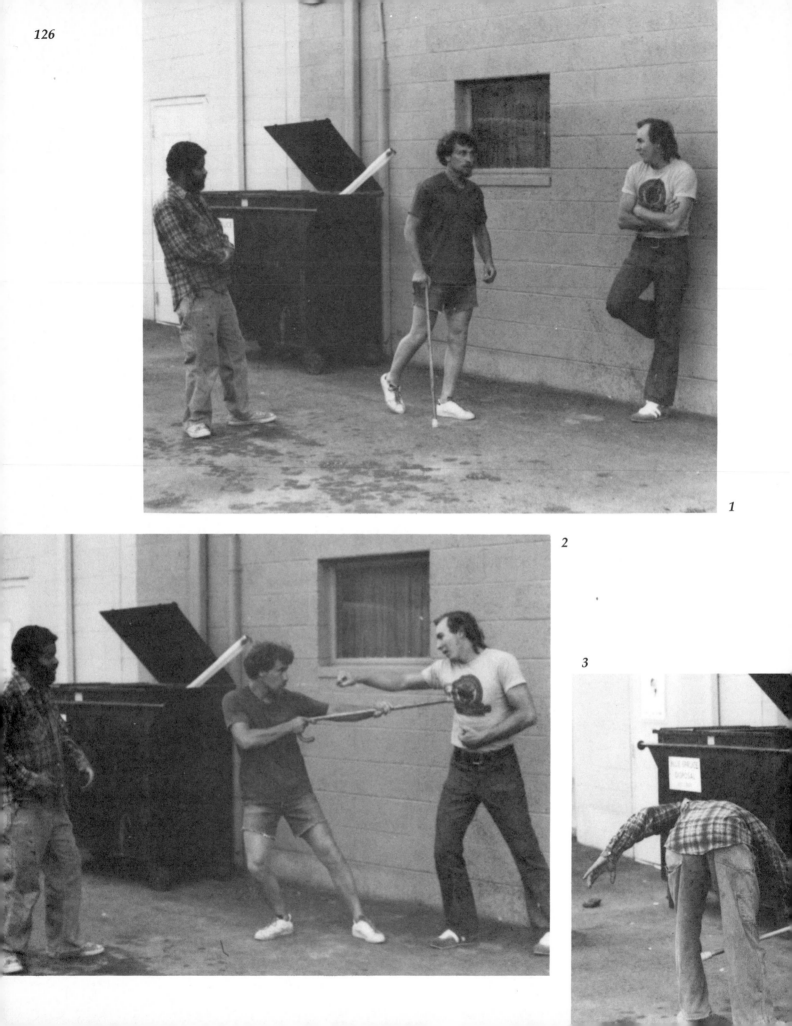

1

2

3

Urban Street Survival I

A soldier of fortune must always be prepared. Even walking the streets of America can challenge the hardiest individual. Hand-to-hand combat skills are an important part of the well-prepared individual's arsenal, that is, if you're interested in survival!

This photo series deals with the use of a common walking stick or cane. It could just as well be an umbrella or any other similar device. This weapon is very useful yet legal in just about every city in the world. It is very important that the "cane" type weapon be thought of as a sword rather than a club. Use it as if it had an edge and a sharp tip.

Man has been using such weapons since the first half-ape picked up a fallen tree branch and smacked his over-bearing next-cave neighbor for trying to steal food. In some parts of the world the use of the fighting stick has been developed to the level of an art. If our anti-gun "law makers" have their way, law abiding U.S. citizens will have need of such skills in lieu of owning a firearm for self defense.

Note that in the photos the defender uses the greater reach of the weapon against his attackers. Also note that he uses a two-hand grip with both palms facing the same way. This grip allows for faster and stronger movements, as well as making it easier to use both ends of the weapon. Always try to strike a bony area like the forearm or knee or wrist and most parts of the head. Shins also make a fine target.

6

5

4

Urban Street Survival II

This series of photos demonstrates the use of a magazine as a defensive (and offensive) weapon. Anything you can lay your hands on becomes an important part of your hand-to-hand fighting.

Left and above: *Two approaches with remarkably similar results.*

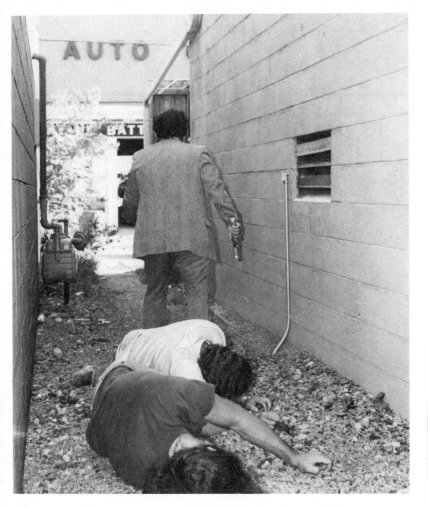

Turning the aggressors momentum against himself, the intended victim strikes at the kidneys.

Left: *In the picture at left the intended victim gains the split second needed to shift his attack to meet his second aggressor.*

Fighting for Keeps

Hand-to-hand combat becomes a bit more tricky when your attacker has a weapon. But even in this case, the well-trained individual can defeat the foe and survive.

An attacking weapon can follow two basic paths—a straight line or a circle, the only two geometric paths possible. Although the two paths may be combined, circular and linear movements are the only trajectories possible.

When dealing with edged weapons, you must remember a knife thrust can be delivered as a stab or a slash, i.e., a linear or circular movement. In a stab, the weapon follows a line, in a slash an arc. When you understand that once your opponent has committed himself to one trajectory—a stab or a slash—you can disarm him by intersecting that trajectory.

If your opponent commits himself through body movement to a straight-line attack, you can predict where the follow-through of that attack will end. Then, by sidestepping that trajectory, you can place yourself in a position not only to avoid but to counter the attack.

Here, as in most fighting-for-keeps techniques, four basic steps hold the key. These steps are:

1. Get out of the line of attack.
2. Redirect the attack.
3. Neutralize the weapon.
4. Neutralize the opponent.

You will have only seconds to predict and react to your opponent's moves. Once he has started his attack, he has committed himself to a follow-through. Judge your distances. React quickly and aggressively. Once your distance is closed, follow through with your technique.

3

2

The Ninja

Ninja: the undisputed masters of unarmed and armed combat. *Ninja:* the very name synonymous with invisibility, stealth, and blackgarbed silent death! A title not easily lived up to!

A joint design of Blackie Collins and Chris McLoughlin, Armament Systems/Product, Unltd., produce the Ninja. The Ninja weighs 4½ ounces. The blade is four inches long and one inch wide. Overall length is eight inches. The blade is hand hollow ground and constructed from 440-C steel. The handle is injection moulded, using "Lexan," which is utilized in bullet proofing systems.

The Gladiator

Many have sought to combine a utility and fighting knife. Some have achieved their goal, but none has taken this desire to the heights which Cordova has reached.

The "Gladiator" is compulsive: one finds his hand drawn toward it. It bedazzles the eye. In hand it flows, becoming a natural extension of the arm.

The "Gladiator" blade is eight inches long; overall length is 12½ inches; weight is 14 ounces with a centered gutter 6/8 of the blade's length on both sides extending from the center of the quillion. It is constructed of one solid piece of 440-C stainless AT 57 Rockwell standard.

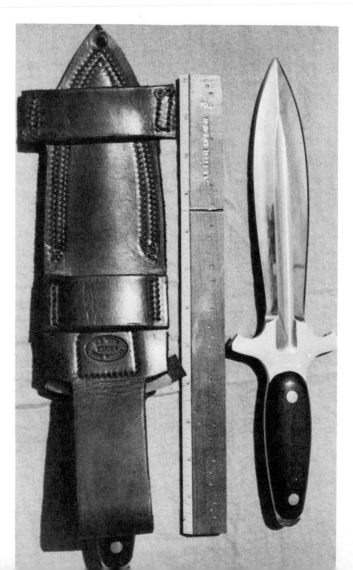

Kitchen Knives

Basically, any kitchen knife, preferably those with a four-inch or longer blade, can be considered a potential weapon. In Hong Kong the favored weapon—because of restrictions on firearms ownership—is the kitchen chopper, a knife-edged cleaver. The housewife is best advised to decide in advance which knife she will pick up if confronted with assault in her home. Essentially, the bigger the better: a large chef's knife is a veritable short sword. It should have a very sharp point and edge. Its blade should not be too flexible: kitchen knife blades may be made in widths from 1/10 to 1/4 inch; the stouter ones are more valuable for thrusting.

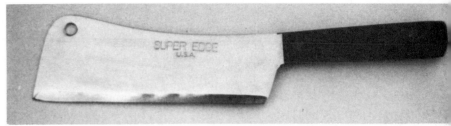

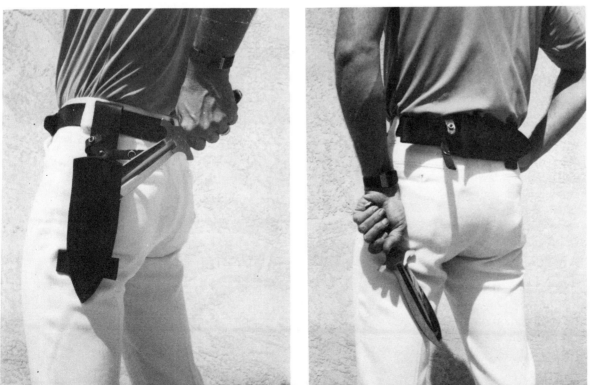

Near left: *With sheath worn transversely, blade can be drawn at an angle.*

Far left: *With sheath worn in conventional manner, blade can also be drawn at an angle.*

Secrets of Modern Battle-Axe Fighting

The innovative soldier of fortune may find any weapon useful. This series of photos demonstrates the effectiveness of a medieval battle-axe. The warrior shown here wields a replica of an 11th century battle-axe, built by Kirby of Eherenburg. Kirby designed it as a Vikingaxe, but it may be no more Viking than English. Throughout the Scandinavian assault on Britain men on both sides fought with what they had on hand. We now think of the two-handed axe as mainly a "British Saxon" weapon, but who really knows? Kirby's axe by any name, is an altogether admirable axe—bright, burly, and battle-worthy.

It weighs 6¾ pounds. The haft is 32 inches long. The blade measures 12 inches around the curve and is 10 inches deep. The point extends seven inches beyond the haft and the hook recurves 3½ inches.

The Kirby axe carries both a pike point and a hook, increasing its versatility. The pike thrust is useful both as the follow up to a cutting stroke or as a tactical continuation after a cut has been delivered. The advantage of the point over the edge—with all of the *armas blancas*—is ease and speed of delivery. Its disadvantage is lack of shock.

The hook may be considered a tertiary device for occasional use as chance may afford, but one should not forget that it is there if needed. Obviously the blade must have been advanced beyond the target for the hook to be used, and that advance ought to have put your man down, but in a melee you may find other targets handy just as you complete a decisive stroke. That hook can jerk a man's leg out from under him, or tear his weapon out of his hands. If a horseman has missed you on the ride past, you can use the hook to haul him out of the saddle.

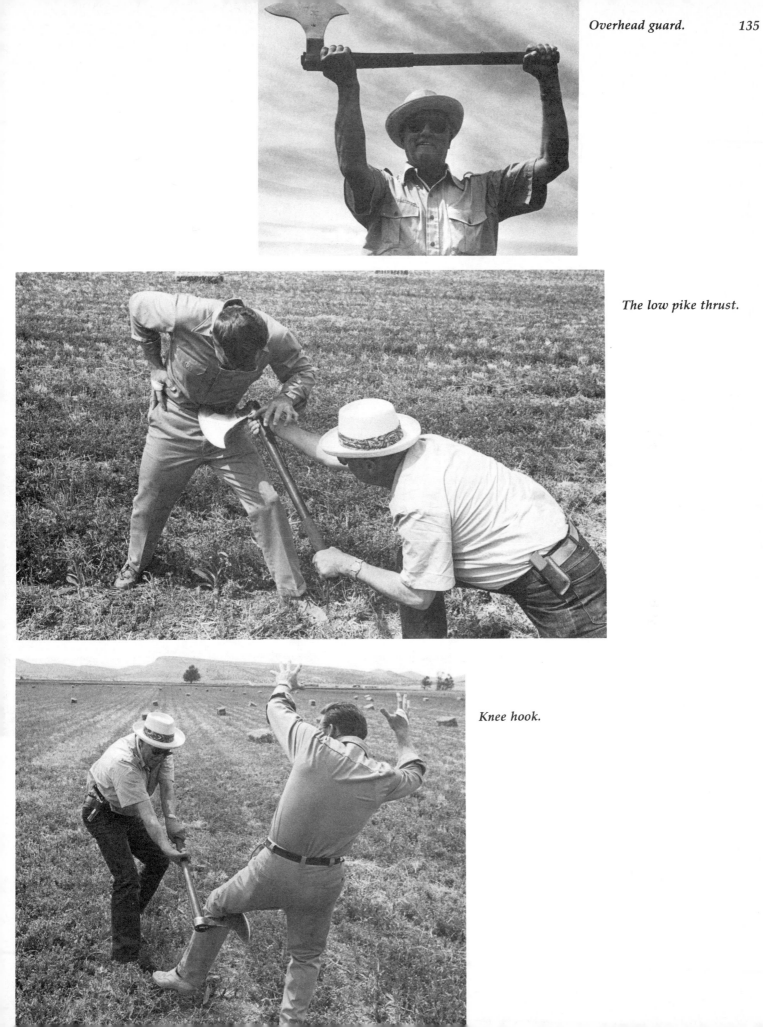

Overhead guard.

The low pike thrust.

Knee hook.

The Crossbow: Whispering Killer

The crossbow, when used within its limitations, is a most effective weapon.

The last time the crossbow was used on a large scale was in the Chinese-Japanese War of 1895. The Chinese used repeating crossbows with poisoned arrows to defend their fortifications.

However, on a much smaller scale, the crossbow has been used in every major and minor war of any duration since then.

In the British War Museum in London there are two crossbows that threw grenades during World War I.

In 1944, the Office of Scientific Research designed two crossbows for underground and commando work. These were rubber powered and came in both a pistol and rifle model. They were code named Big Joe and Little Joe Pedal.

During World War II Major J.M.T.F. "Mad Jack" Churchill of the Manchester Regiment killed a German infantryman at 40 yards with a crossbow during the retreat to Dunkirk.

Montagnard tribesmen used primitive crossbows extensively against the Viet Cong. And VC used primitive but deadly crossbow traps.

The Montagnard tribesmen supposedly followed a marching line of Viet Cong and shot the last man in the line with a crossbow. They shot him through the heart so he would not cry out. They would then pick off the next one up the line.

There is no shock to speak of with an arrow. It kills by hemorrhage or puncturing vital organs. The only shot that I know that ever brought any type of large animal down in its tracks is a spine shot.

You may be anticipating a head shot. As a rule this is an inadequate shot. The human skull has adapted to having slow-moving projectiles bounce off. Even though a direct hit in the center of the skull would pass through, the size of the kill area is limited to about three-by-three inches.

To psychologically harass the enemy, the shooter may want to use poisoned arrows.

CZ 75: A Good Communist Pistol

Admired by all, imitated by some and previously thought attainable only at exorbitant cost, the much-coveted CZ (Czech) 75 pistol is actually readily available to those willing to untie some red tape.

Designed by the brothers Josef and Frantisek Koucky at the Ceskoslovenska Zbrojovka, Narodni Podnik in Brno, the vz. (model) 75 is a successful blend of innovation and the best features of several other well-known pistols.

The CZ 75 owes little to its immediate predecessor, the Model 52 pistol, which operates with a locking system taken from the German MG42 GPMG and is chambered for a more potent loading of the ComBloc 7.62x25mm bottleneck cartridge. Chambered in 9mm Parabellum, the Model 75 does not seem to have been engineered for the Czech military, but for Western and Third World police and military markets. That's not really surprising, since the Czechs have been major arms merchants to the world since the inception of their nation in 1918.

But, can you obtain this communist Cadillac for anything less than the $1,100+ scalper's price? Yes indeed. Brno firearms are imported into Canada by Pragotrade (Dept. SOF, 307 Humberline Drive, Rexdale, Ontario, M9W 5V1, Canada). Their price for the CZ 75 is $390 (U.S.) which includes an extra magazine and plastic grip panels. The military version, with phosphate (Parkerized) finish, a Colt Commander-style hammer, lanyard ring and wood grips is not available from Pragotrade. Walnut military grip panels are $20 extra if ordered with the pistol. Additional magazines are $24 each. That's the good news. The bad news is that Czechoslovakia has "Unfavored Nation" status with the United States and their products are subject to a 55-percent duty.

To purchase a CZ 75 from Pragotrade, a licensed firearms dealer must write to the BATF (Bureau of Alcohol, Tobacco and Firearms, Washington, DC 20226) and obtain an ATF Form 6, which is an Application and Permit for Importation of Firearms, Ammunition and Implements of War. It takes about three to six weeks for the Form 6 to be processed and approved after return to the BATF. All the form's questions are self-evident, except that paragraph 8g, Serial No., should be answered, "to be submitted on Form 6A."

When it has been approved, the Form 6 will be returned to the dealer with two copies of ATF Form 6A. Xerox the approved Form 6 and send it along with your payment to Pragotrade. Within two weeks the pistol will be shipped (by air freight, F.O.B. Ontario) to your local U.S. Customs branch. The dealer must then clear U.S. Customs by properly executing the Form 6A and paying the duty. He then sends his copy of Form 6A back to the BATF and that's the end of the matter. The total cost will come to about $624 plus shipping charges. At this price one of the world's primo 9mm pistols becomes a palatable proposition.

P-38K: Short Barrelled Dynamite

The Walther P-38K is brand new in design, size, and intent, yet familiar for many reasons. The basic Walther P-38 has a fine reputation as a service pistol, was the winner in the pistol trials held in Germany prior to World War II, and was the choice of the West German Army after the conflict. Like all Walther pistols, it shoots where you point it, functions reliably under adverse conditions, and is as safe to operate as it is easy to field strip.

The P-38K's most obvious difference from the standard P-38 is its substantially shortened barrel. Though the most apparent feature, it is one of the least important. Most of us have seen P-38s with shortened barrels. On occasion, German undercover agents used such snubbed pistols during World War II, and the popular TV shows, "I Spy" and "The Man From UNCLE," both featured a short-barreled P-38. In the U.S., one in three gunshops has a homemade hacksawed P-38 in the used-gun showcase.

Hand Cannons

Handgun ammunition made the transition from lead to jacketed high-performance ammo beginning in the mid-'60s. Unfortunately, handgun bullet development for hunting seems to be suffering from a severe case of arrested development that began in the early '70s. In a 10-year period, fairly decent progress was made—since then everyone seems to be sitting on their laurels waiting for something to happen.

Well, it's happened! SSK Industries, Rt. 1, Della Drive, Bloomingdale, OH 43910 (614) 264-0176, has put some real power in a pistol. The Thompson Center Arms Contender (T/C) has a strong, proven action. SSK has developed two cartridges for use in their custom T/C barrels, each of which easily exceeds a ton of muzzle energy per shot.

The .45-70 has been around since 1873 and has been highly developed as a rifle cartridge. Its reputation as a killer is well deserved. The other is new—the .375 JDJ. It's a full-length .444 Marlin case necked to .375 with a 25-degree shoulder and one caliber neck.

No other conventional handgun—excluding bolt-action handguns—can approach the power that can be packed into a T/C.

The Auto Mag Mini Sniper System

International Pistol Champion Lee E. Jurras built a limited edition of this exotic handgun. In a letter (4/12/75) to Soldier of Fortune *Publisher Bob Brown, he described his specially designed handgun.*

"The entire unit consists of the highly modified 8½" barreled .357 AMP, scoped, two magazines, clip depressor, box of ammunition and detachable shoulder stock, all packed in a custom compartmented Belting leather attache-type case. The purchasers' initials are burned in and the combination lock is set for his birthday. In other words a very personalized item. There will be 25 only of these units built to these specifications with serial number LEJ-01 being kept by myself and serial number LEJ-02 will be sent to Italy to be very elaborately engraved. The balance of the 23 will be sold to the discriminating sportsman throughout the world.

"For the serious sheep hunter, who enjoys the hunt and stalk, this has to be probably the world's lightest combination for sheep hunting, where weight is the all important item above the ten thousand foot level; or for the international sportsman who might care to engage in a bit of chamois hunting while on a quick business trip to Europe, yet does not want to be encumbered with a bulky rifle.

"We are not announcing or discussing price on the L.E. JURRAS CUSTOM MODEL 200/INTERNATIONAL; it's priced and shown by appointment only. Color photos are available to individuals overseas, and at the time of this writing, there are approximately 13 guns left for sale. This exotic item was designed for the affluent sportsman and collector only, and unless one wants to discuss prices in the middle four figure range, we shan't waste one another's time."

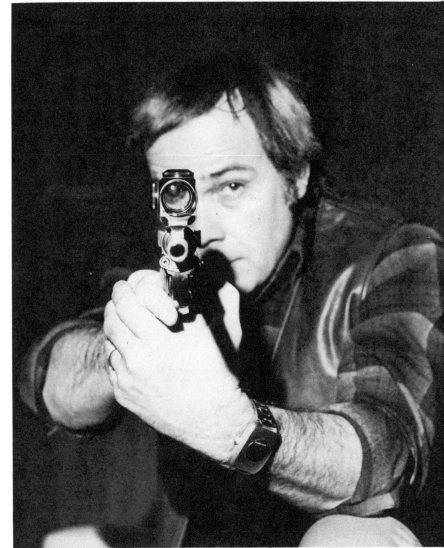

Left, above: *With proper training anyone can fire weapon competently.*
Far left: *Allan Bateman busts a gallon of gas with 220 grain .375 JDJ.*
Left: *Hand cannons require special firing techniques because of their tremendous recoil.*

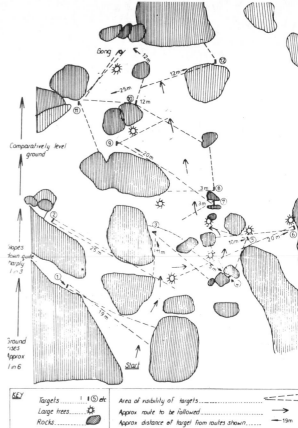

Shooting for Free

Practice makes perfect and, when it's either you or the other guy, you can't afford not to be perfect. Participation in pistol competitions and combat simulation courses can keep your skills finely tuned.

Above: *Colt .45 at ready, Ross Seyfried comes over the wall on the assault course during the International Practical Shooting Confederation's (IPSC) competition held in June 1981.*

Right: *Edith Almeida shoots her way to 12th overall in the World Practical Pistol Championships held in the Transvaal.*

Right: *Young lady going full auto. She had never fired a firearm before this picture was taken.*

Left: *Example of a simulated jungle course where the shooter will be ambushed at numerous points. This course was used in the World Practical Pistol Championship Shoot held in Salisbury, Rhodesia, August 1977.*

Bottom: *John Shaw fires Clark custom-modified Colt .45 from "Rhodesian" wall. In 1981 John won his second United States championship — the only shooter ever to do so.*

Top: *One of the latest versions of the AK produced by South Africa, named the R-4. The AK was originally patterned after the German MP-40.*

Above: *Rare photo of AK equipped with silencer.*

Right: *AK used for home defense in Rhodesia.*

The AK-47: Soviet Killer

The reader needs no introduction to the Soviet AK-47. Hardly a day goes by without one being seen in the hands of some soldier or guerrilla on television. What you may not know, however, is that the development of the AK-47 can be considered an historical accident and that the rifle is produced in Communist and non-Communist countries alike in a variety of forms other than the Soviet model.

The popularity the AK-47 now enjoys around the world is not the result of a conscious effort on the part of the Soviet Union to develop an assault rifle. It is the result of the combination of a reliable design, imposed standardization in Eastern Europe, and massive Soviet giveaways in the Third World.

At least 11 countries manufacture some form of the AK-47, and many more issue them, making it the most common of all assault rifles now in use. While some experts point out that the Soviet M1943 cartridge is not a particularly well designed combat round and that the AK has some design flaws, it fulfills the two prerequisites of any infantry rifle: it is reliable and it kills.

Right: *The AK is captured and recaptured in war-torn Africa.*

The AR-15: Black Gun Blaster

The M-16/AR-15 "black gun," which is our main battlefield small arm, is not known for its accuracy or tons of energy. It's not even legal for hunting in most states since it is so underpowered—some battlefield weapon, huh? To be very honest, it's worthless—or is it?

There is a gunsmith in Boulder, Colorado, named Mark Chanlynn, who shoots as an N.R.A. expert in high power competition with an AR-15! Although this may not impress you, it should. One phase of these matches is fired with iron sights at 600 yards. Mark won the Colorado State 600-Yard phase with this rifle in 1977.

Mark takes an AR-15, goes through it, and produces a rifle capable of three-inch groups at 600 yards.

This is obviously an excellent target rifle since it shoots in bolt action rifle matches (it is not in the service rifle class). This gives it an excellent advantage in rapid fire competition: low recoil and excellent accuracy. The potential for special unit or SWAT application is fantastic. As a counter-sniper weapon, it can shoot with pinpoint accuracy and yet allows for rapid follow-up shots. Most SWAT teams deploy with a bolt gun sniper and an AR-15 or M-16 backup. Now the same job can be done by one man if the situation allows, or you could have two semi-auto "accurate" snipers.

What then is Mark's AR-15 all about and what are its advantages? First, it is sold with the customer's choice of barrel, whether by Hart, Douglass, or whomever. The rifle weighs up to 11½ pounds, due to the heavy 24-inch barrel and steel forearm. About 1½ pounds can be cut off by using an aluminum forearm. The 10 to 11-pound weight allows for more shooting stability and lower recoil. Mark can't report on any malfunctions in 1700 rounds of firing—so it appears not to be temperamental.

The M-16 straight line stock design with a pistol grip further reduces recoil and adds to a more natural point of aim. This stock also allows for both right or left-hand use—a definite advantage over most bolt rifles. It has the lowered-training-time advantage for SWAT units, since they usually train with AR-15s or M-16s.

The Sidewinder SMG: The Ultimate Automatic Weapon

The Sidewinder SMG was first designed and built from scratch by Sid McQueen, head of S&S Arms Co. Chuck Taylor filed this report in October of 1979 after test firing this weapon with members of the SOF staff:

When I first met McQueen and his staff at SOF Publisher Robert K. Brown's house, we discussed the probable solution of warfare and its effect upon the standard infantry squad as we know it. Our consensus was that the art of war is becoming so efficient that, very probably, within a very, very short time period in combat, almost all conventional command structure will cease to exist because of high personnel and equipment losses. This being the case, the infantry squad as we know it will become instead a "survival squad."

Projected size of the "survival squad" is three to five men. Any group of personnel larger than this would bring swift artillery, aircraft, or other enemy response. During our discussion, we decided the SMG, a grenade launcher and an anti-tank weapon, preferably of the shoot-and-throw-away type, and the telescope-sighted rifle would be the most useful weapons for such a unit. Why? Well, future battlefields will usually be in urban areas. Recent events in the Middle East are examples of this emerging pattern.

The battle rifle, assault rifle, light machine gun, and squad automatic rifle will see little use in future urban warfare since they require organized training for potential operators to attain any degree of efficiency. Also, these weapons are not suitable for close in fighting and they require logistical support that is out of the question in such a scenario.

The SMG, on the other hand, requires less training for operators to utilize it efficiently at close combat ranges. Because the SMG is cheap enough to allow its abandonment, if necessary, under tactical circumstances, its operator can take the weapon, equipment, ammo, and food of the adversary whom he has just killed if he is low or out of ammunition.

As fighting continues, the most seasoned squad man will most likely use the telescoped rifle, as he will be the most skilled. Remember, almost all participants will be relatively unskilled initially. The man who survives long enough to be senior will do so because he is smartest, quickest and most ruthless. It doesn't take great "skill" as we now know it to effectively use a 'scoped rifle against a human target at ranges inside 200 meters.

The AKS-74 Soviet Assault Rifle

AK-74 Specifications

CALIBER: 5.45x39mm
MUZZLE VELOCITY: ....2,950 fps
WEIGHT, empty:......................7.9 lbs.
LENGTH (AKS variant): overall 36.5 inches; with stock folded 27.25 inches
BARREL: Length 18.5 inches; chamber and bore chrome-plated 4-groove, right-hand twist, 1 turn in 5.8 inches
FEED: 30-round plastic magazine, will also accept 40-round RPK-74 SAW magazine
SIGHTS: Front................adjustable post
        Rear ....tangent, U-notch, elevation to 1,000 meters
OPERATION: ....Gas, no regulator, selective fire
CYCLIC RATE:..................600-650 rpm
METHOD OF LOCKING: ....two-lug rotary bolt
MANUFACTURER:...unidentified Soviet arsenals

Above: *Field-stripped AKS-74.*

Above: *The AKS-74 receiver group. Center position is marked "AB" for full auto; lower position marked "OA" for semi-automatic.*

Right: *AKS-74 with stock extended. Note typical AKM finger swells on lower handguard.*

Left: *This new copy of the Russian AKM produced by the Red Chinese is simpler and cheaper.*

Right: *Top to bottom, MKG semi-auto, full-suppressed MK5 SMG, and two MK4s.*

Right: *Top to bottom, the new Sterling semi-auto MK VII 4″ and 7.8 barrels and the full-auto versions with vertical foregrips.*

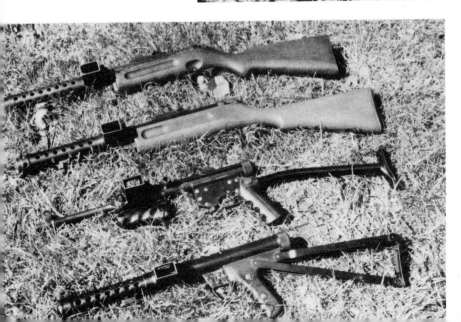

Sterling Gold

Touring Sterling Armament in Dagenham, Essex, England, is well nigh a trip into the twilight zone of small-arms manufacture. Sterling submachine guns, every bit as good as their stamped-sheet-metal tin-can-gun counterparts, are still manufactured by methods approximating British battleship engineering.

Left: *Lanchester Lineage, Hugo Schmeisser's MP28II, pirated Lanchester MKI and two Lanchester prototypes.*

Below: *The Lewis gun which the Bren, at right, replaced.*

The Unbeatable Bren

The Bren has served the British Army well for almost half a century now. It has stoutly helped hold the line on the sands of North Africa, in jungles from Burma to Malaysia, against the Mau Mau in Kenya, on the frozen hills of Korea and, most recently, in the Falkland Islands.

Proven in the crucible of war for more than four decades, the Bren needs no test and evaluation from any weapons technician. It is the best magazine-fed light machine gun ever-produced. It will continue to be that far into the future.

The .45 Caliber Thompson

The Deadliest Weapon, Pound for Pound, Ever Devised by Man—Time Magazine, 1939.

The above was, and still is, a most appropriate definition of the (in)famous Thompson submachine gun. The "Thompson," as it has since become known, has chattered its way into the annals of American folklore as an artifact of both cultural and military history.

Initially conceived by General John Taliaferro Thompson as a military weapon during the "Great War" of 1914-1918, the TSMG was intended as a "trench broom" for clearing trenches and other fortifications of that era. At that time, the machine gun in its true form, i.e. with tripod, belt feed, traverse and elevation mechanism etc., was a new innovation in terms of military deployment and tactics, although it had existed quite a number of years in reality. The use of fully automatic weaponry in tactical ground combat caused the methods and strategy of the day to become obsolete almost overnight and made World War I the most gruesome war in history to date.

Eventually, the Thompson found its first place in American society in the hands of the gangs and "beer wars" of the late '20s. This, much to dismay of the highly idealistic General Thompson, became the *era* of the Thompson. Although grossly overpublicized, the Thompson submachine gun *had* been used in several gangland killings, the most spectacular of which was the so-called "Saint Valentine's Day Massacre." This seemed to place the TSMG on the wrong side of the highly-publicized motto of Auto-Ordnance: "On the Side of Law and Order!" Needless to say, civilian sales were not stimulated by such goings on.

The Thompson, now called, "Tommy Gun," "Chopper," "Chicago Typewriter," "Piano," and Squirtgun," was even further maligned through its use by depression-era bandits John Dillinger, "Pretty Boy" Floyd, Clyde Barrow and Bonnie Parker, "Babyface" Nelson, "Machine gun" Kelley, and Arizona "Ma" Barker! Their exploits, much publicized in the news media, catapulted the Thompson to the status of being an underworld symbol, causing even more heartache to General Thompson.

In 1939, the threat of world war and the distressing realization that the SMG would play a major part in any modern conflict jolted the U.S. Government out of its isolationist euphoria of the '20s and '30s and caused the military community to begin looking for a suitable SMG. Except for limited use by the U.S. Marine Corps in Nicaragua in 1928, and the U.S. Army Cavalry, the TSMG had been virtually ignored by the military in spite of the continuous efforts of Marcellus Thompson and the Auto-Ordnance Corp. staff. Finally, someone in the Army Ordnance Dept. remembered the TSMG and arranged for it to be tested again. It passed Army tests with flying colors and was adopted in its M1928 (U.S. Navy) version and began immediate mass production for issue to the U.S. military forces and all U.S. allies!

Facing page, left: *The Bren series, front to rear, L4A2 in 7.62 NATO, MKII Canadian Inglis in the MKI 7.92mm configuration, early MKI in .303 British.*

Military Material

The mercenary: Whether accepted as a soldier or rejected as an entrepreneur of death really doesn't matter. Throughout history, the mercenary has always played a role in the global drama of good and evil. Those who play this role and adopt its lifestyle do so of their own choice. Those who play, play to win. The intent of the following information is to provide some insights concerning equipment that can increase the options and survival odds in the mercenary's favor.

Above, left: *The boot on the right is a good design; the one on the left should be used for recreational hiking.*

Left: *North Face geodesic-design tent.*

Below: *Tropical tent by Sierra West.*

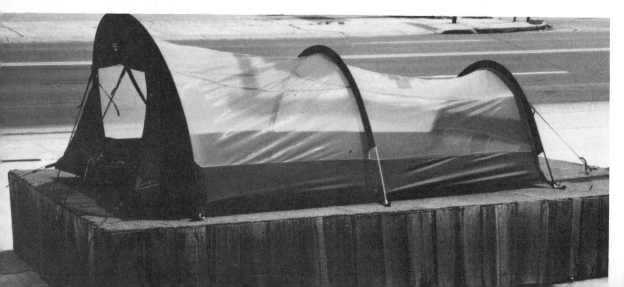

Right: *Kelty® Tioga bag and frame.*

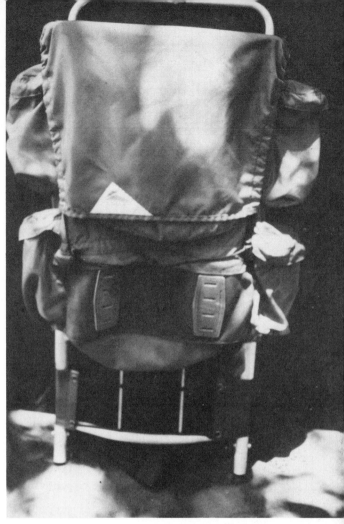

Left: *Camouflage is a must. Most companies offer their wares both cammy and recreational colors.*

Below: *The most effective sleeping bag shapes. On the right is a modified mummy; on the left, the standard mummy design.*

Left: *The Fitzroy from Trailwise is an excellent example of a freestanding A-frame tent.*

Bullet-proof Vests: A Matter of Life

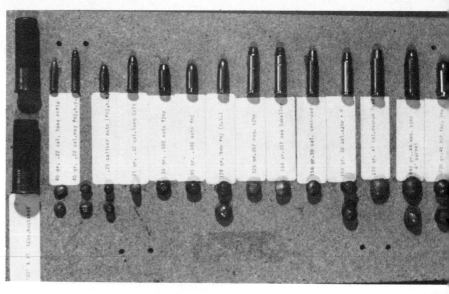

If you don't have it on when trouble comes, you bought it to be buried in.

Most fatal gunfights in the United States occur from 12 feet or closer and are over in an average of less than three seconds.

Soft, lightweight, bullet-resistant vests are concealable enough today to offer real protection to police, bodyguards and security personnel, but to be effective you have to be wearing them when trouble comes.

In addition, to get your money's worth from soft body armor you need facts before buying, especially as many vest salesmen will mislead you.

You must first decide why you need it, how often it will be worn, what coverage you want and what it will have to stop.

To decide what threat level you wish to stop, canvass your area to check the types of confiscated weapons and types of weapons most likely to be used against you. You must be realistic and get a vest that you will be able to wear daily. Threat levels are listed according to the types of projectiles the armor will have to stop:

Threat Level I—22 LR (40-gr.); 38 spec. RNSP; 12-ga. No. 4 shot.

Threat Level IIA—9mm FMJ (1090 f/s) 124-gr.; 357 mag. 158-gr. lead.

Threat Level II—9 mm FMJ Remington 124-gr.; 1175 f/s; 357 mag. JSP Speer 158-gr.

Threat Level III—7.62mm (.308 Winchester) 150-gr.

Threat Level IV—.30-06 military APM-2, 166-gr., 2750 f/s.

You should start by considering a Threat Level IIA vest, since most U.S. handgun rounds in street use can be stopped by this vest. In Europe, you will need a Threat Level II vest, since there are numerous 9mms with high-velocity, steel-cored bullets. A Threat Level II will also stop most 9mm submachine-gun rounds.

Above: *Front view showing Ken Pence of the Nashville P.D., wearing vest properly so that it doesn't show.*

Top: *Bullets fired from three feet that did not penetrate a standard 15-layer Kevlar 29 Vest.*

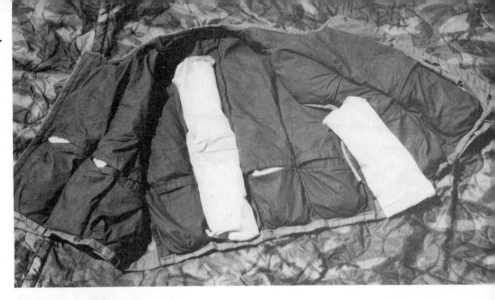

Right: *Inner liner of a Soviet flak-jacket. Note the exposed plastic bags filled with cotton wadding.*

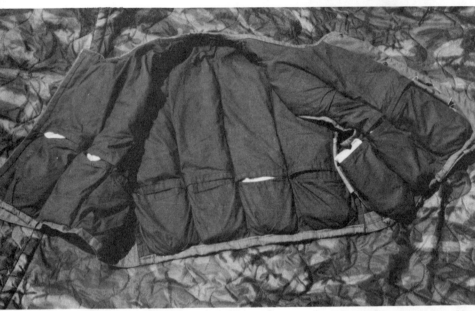

Left: *Elongated pockets hold a single heat-sealed plastic packet stuffed with partially processed cotton wadding.*

Below: *Front view of Soviet body armor. Overlap covers a row of four eyelet-and-hook steel fasteners, thus doubling protection for a five-inch strip running from the clavicle to the waist.*

Soviet Body Armor

SOF was recently presented with the unique opportunity to examine in detail and evaluate the only known specimen of current-issue Soviet infantry body armor in the West. More appropriately called a "flak-jacket," this particular vest was captured during Operation Protea (see "To Russia With Love," SOF, January '82) by South African Army personnel. It was removed from a Russian Army adviser taken prisoner in the raid.

And what will this lightweight three-pound Soviet vest stop? Very little, if anything, I'm happy to say.

Several important inferences may be drawn from this examination of current Soviet body armor. But most importantly, its poor quality and lack of adequate protection are fully consistent with Marxist ideology's low regard for the individual, whose protective garments consequently have low priority in overall Soviet military research and development.

Close-up of Rollei XF-35.

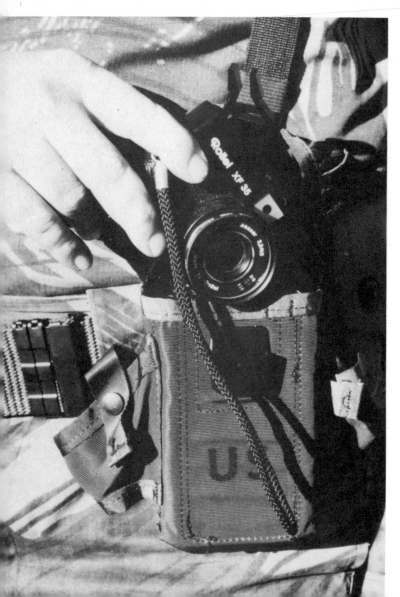

Above: *Olympus XA with its case pulled apart, ready for use.*

Left: *A compact camera can fit in an ammo pouch, protected and out of the way.*

Above: *Olympus XA with the flash attached.*

The Hidden Camera Trick

A camera is a useful tool in military missions and police work. Crime-scene photography, evidence documentation and surveillance photography have been a part of law enforcement for many years.

A compact camera is especially useful in military intelligence, particularly when bulk and weight are considerations. Photographic documentation provides more information and better confirmation than purely verbal means. The imagery is also easily reproduced as many times as necessary for dissemination and interpretation. A recon unit equipped with a camera can increase its information-gathering capability several times.

If you don't want the evidence for later use, consider binoculars. They can do three things: make objects look closer, make objects brighter, and look good hanging around your neck.

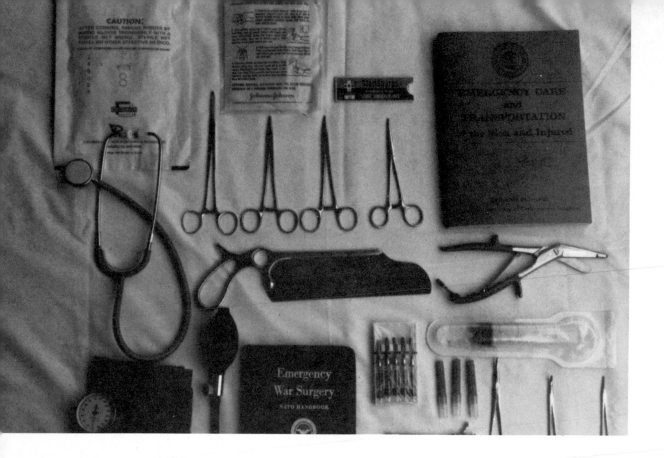

First Aid for the Merc

Congratulations! You've just landed that big contract and now you're off to Asia, Africa, one of the Arab countries or perhaps South America. You're looking forward to a new adventure, but as you clear up last-minute obligations and prepare to leave for the airport, take a minute to think about medical care and personal health. Wherever you are going, odds are that the area won't have a health-care delivery system which measures up to what you've come to expect in North America or Western Europe. Depending on just where it is in the world that you're heading you can expect to encounter communicable diseases seen only infrequently in the West, exotic and perhaps not palatable foods and "doctors" who may be little more than orderlies.

Staying in good health will depend very much on you, and your chances of living to spend that bonus increase with your knowledge of elementary medical care principles. The time to think of these aspects of your lifestyle is now, while you are still in a Western country, not when you are stuck in the bush with the nearest hospital three days' hard slog away.

By the way of illustration, your personal kit should contain:

   1 pair of needle-nosed tweezers
 25 band-aids (Johnson & Johnson plastic strip)

1 Cutter snake bite kit
1 oral thermometer, with case
1 tube of betadine ointment
1 container of "Emprin," "Tylenol" or other pain reliever
2 ampules of ammonia inhalant
1 pair of small scissors
1 pair of forceps or a hemostat
1 Benzedrex inhaler
1 tube of chap-stick
8 sterile packaged alcohol swabs
4 antiseptic wet cleansing towel packs
1 container of 50 mg. Demerol tablets
1 container of 25 mg. Bonine tablets
1 container of 50 mg. Lomotil tablets
1 roll of 1" tape
1 small bottle of eye wash
1 packet of sterile suturing needles and sutures
4 2"x2" 8-ply sterile gauze sponges
4 3"x9" nitrofurazone sterile gauze pads
4 4"x4" sterile cover sponges
1 book of matches
1 pack of sterile surgical blades
4-6 safety pins
1 roll of gauze bandage.

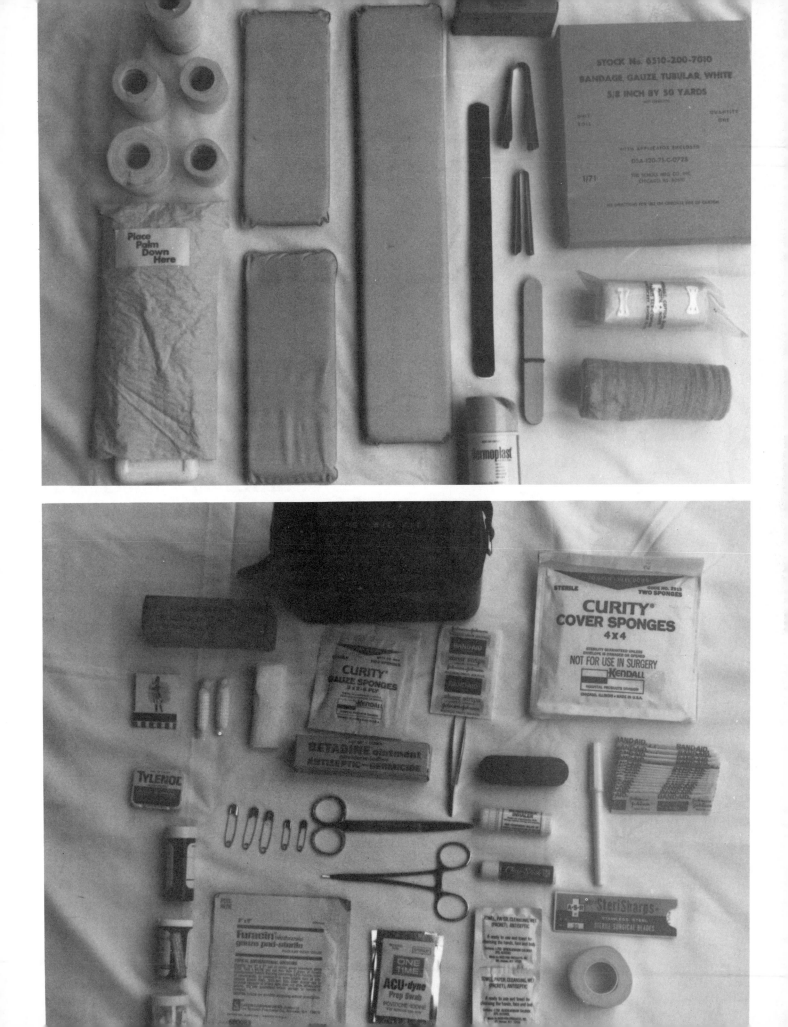

Inside the World of Soldier of Fortune Magazine

FACES OF WAR

Top left: *Bob Brown, publisher of* Soldier of Fortune *magazine, on assignment in South West Africa with SOF's answer to pet control.*

Top right: *Maj. Thomarat of the Thai Rangers, 4th Battalion, the Black Panthers.*

Bottom left: *U.S. Marine Recon in the bush in Puerto Rico, armed with an M-16 A1.*

Right: *In the latter stages of the Zimbabwe/Rhodesia conflict, everyone was required to do his part. Old-timers like this fellow brought years of experience into the bush.*

Right: *Everything is a little easier after the rains of the monsoon season, but the tension on the border is still there.*

RANGE WAR IN SOUTH AFRICA

Ovambo troops ride with white South Africans on a mounted patrol in Namibia/South West Africa. The rifle grenades are carried as a break-contact weapon in case the troopers ride into an ambush.

Below: *Between Angola and South West Africa is a no man's land of patrol and ambush.*

Left: *In hot pursuit of S.W.A.P.O. Mounted units are used primarily to track guerrillas; the horses are too valuable to actually ride into a fire fight.*

Staffer John Donovan makes contact with other members of the SOF team during ambush and counter-ambush training with El Salvador's Atlacatl (Immediate Reaction) battalion. In August 1983 SOF sponsored a three-week program that included training in medical, explosives, airborne, maintenance and chopper gunning. This was not the first time SOF sent a training team to a hot spot, or to El Salvador for that matter.

ON THE FRONT LINE OF FREEDOM

Above: *Shot through the foot, this taken, guerrillas opened fire as the*

CENTRAL AMERICA

Signaling a safe L.Z. in El Salvador — calling in the choppers.

Guatemalan trooper is rushed toward a waiting Evac unit. Moments after this photo was chopper cleared the L.Z.

BEATING THE BEAR

Burning up the mountains on a sunny afternoon with a Chinese-built ZPU-1 14.5mm AA gun. "Make my day!"

Right: *SOF publisher Bob Brown waving an assault rifle while "going native" to get the inside scoop on the siege of a Soviet fire base by Afghan freedom fighters.*

IN AFGHANISTAN

Top: *SOF staffer Dave Isby displays a Soviet AKR while on assignment in Afghanistan.*

Bottom: *Rebels in training. Experts in mountain warfare, the Afghans have forced the Soviets into a road-bound mentality, a strategy for defeat.*

ON THE FIRING LINE

Below: *Gurkhas blasting away while training in Belize. Since 1948 this brigade from Nepal has been a regular unit in the British army and are not mercenaries as such. They are rotated where ever there is a potential powder keg. After they arrive, things usually calm down.*

Above: *On patrol in Ulster. The Royal Regiment of Wales is one of the line units patrolling Northern Ireland.*

Above: *Firing the Ultimax 100 5.56 NATO SAW at the American Defense Preparedness Association meeting in Quantico, Va., October 1983.*

Above: *Mohammad Kareem, former brigadier in the Afghan army, fires the Soviet AKS-74. He is now a Mujahideen camp commander.*

Right: *Going full auto on the test range at the Third Annual SOF Convention in Charlotte, N.C., in October 1982.*

After searching for years, the SOF team in Afghanistan finally track-ed down the elusive—and awesome—Russian AGS-17 automatic 30mm grenade launcher. SOF Military Small Arms Editors Peter G. Kokalis and Jim Coyne were the first Americans to test fire this Soviet killer. SOF has consistently sought out, found, test fired, and published photo-graphic evidence of Soviet weaponry. For the rest of us, there is no other way to find out about these weapons—unless we want to travel 12,000 miles to Peshawar, Pakistan, then smuggle into Darra Adam Khel, and pay a local (and shady) gun dealer $45,000 to buy an AGS-17. But, even then, we'd still have to figure some way to get the damned thing back into the United States.

SOF Convention Collage: The Merry Band Gathers

On 12 through 17 October 1982, we were at it again—for the third time. SOFers from across the country and around the world came to Charlotte, N.C., to watch, look, listen and take part in the SOF Third Annual Convention and Three-Gun International shootout. Spread out across the North Carolina countryside, conventioneers took in the gun show at the Charlotte Civic Center, the shooting match, factory demonstrations and automatic weapons exhibition at the Charlotte Rifle and Pistol Club Range, airborne operations with the 1st Airborne Division at the Lancaster County Airport

Drop Zone, and the various seminars, movies and war-stories-in-the-bar at the Holiday Inn headquarters hotel.

Saturday night's banquet and beer bust at the Civic Center culminated 1982 convention activities, with a Southern-style pig roast followed by words of wisdom and encouragement from guest speakers—including G. Gordon Liddy, Gens. Heine Aderholt, John Singlaub and William C. Westmoreland. To top it all off, the banquet concluded with the marriage of SOF's Demolitions Editor John Donovan to his lovely bride, Pam.

Our readers got together once again, and, as usual, rekindled old friendships, swapped new stories (and some old ones) and had a great time.

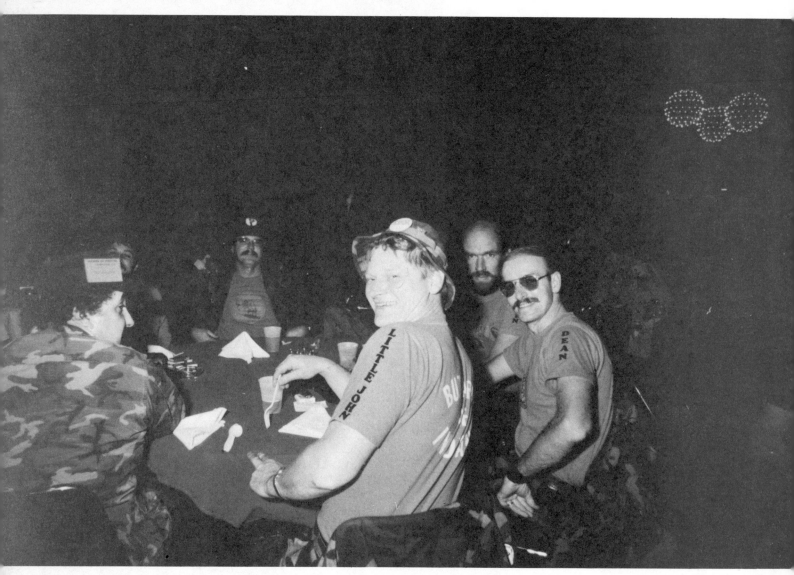

Top: *Color guard courtesy of the Charlotte Marine Corps Reserve Unit.*

Bottom: *Buzzards of Indiana can't wait to inhale the barbecued pig.*

Right: *Bob Brown talks to the press.*

Below: *In salute to Project Freedom, the Bull Simmons Memorial Award was presented to two Vietnam vet ex-POWs.*

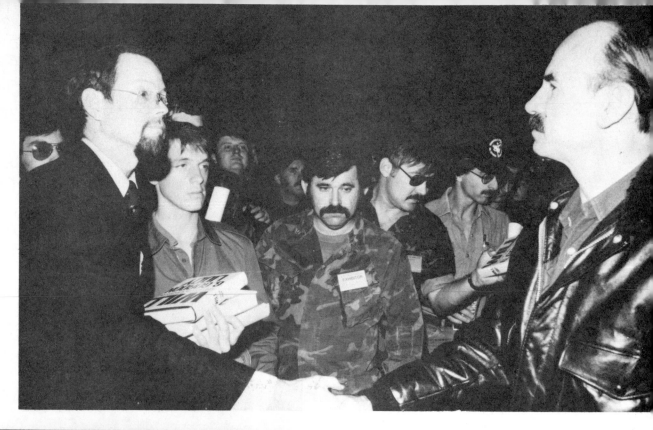

Right: *G. Gordon Liddy signs his book, makes new friends at the gun show.*

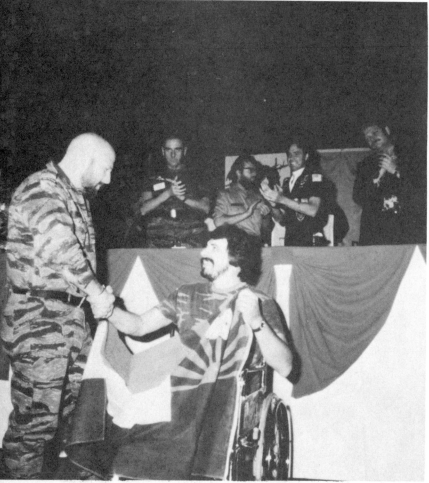

Above: *Vietnam vet Mel Tatrow won the raffle for the Karen Liberation Movement flag, symbol of freedom in Burma.*

Gen. William Westmoreland (right) addresses 1,200 conventioneers at the banquet, and below adds his name to the Viet vet's painting which will be on permanent display at the Smithsonian Institute in Washington, D.C.

On target! Demo jump in front of headquarters hotel.

Above: *Soldier of Luck Capt. Larry Dring gives his "Man Against Tank" seminar, using visual teaching aids.*

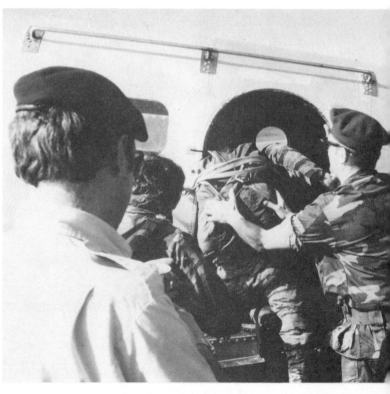

Right and facing page: *1st Airborne Division ops were in high gear for the Third Annual Convention, with 111 jumpers going out the door for the first time. A total of 173 operational jumps were made, including some sport jumping and a PIAD demo from a C-45.*

Sheriff's officer fires H&K MPV SD SMG.

Springfield Armory's rock-n-roll demo.

Right: *About 500 conventioneers were treated to a seven-hour automatic weapons demonstration at the Charlotte Rifle and Pistol Club range.*

A pyrotechnic target goes up.

Below: *Lady shooter blasts away on the shotgun course.*

Facing page, top: *Representative from Lanchester, U.S.A. firing Sterling MK 4 SMG in 9mm Parabellum. Next to him two more reps are firing the Sterling MK 5 silenced SMG also in 9mm Parabellum.*

Left: *Almost 120 top shooters competed in Three-Gun International Match.*

Below: *Ken Hackathorn opening up with a MK 1 Bren gun in .303, vintage 1940.*

These advertisements and bounties have all appeared on the pages of SOF over the past ten years. The offer on Idi Amin still stands.

In addition SOF will pay $1 million to any pilot, trainer or crew member to defect with a working model of a Soviet Hind helicopter.

Support
AFGHAN FREEDOM FIGHTERS

SUPPORT THE BRAVE PEOPLE OF AFGHANISTAN IN THEIR FIGHT FOR FREEDOM AGAINST SOVIET AGGRESSION AND OCCUPATION.

All funds collected will be donated to pro-western Afghan resistance groups selected by the SOF staff.
These funds will be used solely for the purchase of arms, ammunition and medical supplies as specified by the groups receiving assistance. No funds collected will be expended for salaries or administration.

Donations are NOT tax deductible.
Send your donation to:
Afghan Freedom Fighters' Fund
Box 693
Boulder, CO 80306

WANTED

IDI AMIN
FOR NUMEROUS CRIMES AGAINST HUMANITY
$10,000.00 REWARD
IN GOLD

For Information Leading to the Capture Alive by Proper Authorities and Delivery to Uganda For Trial

Reward Offered By
SOLDIER OF FORTUNE
Magazine

Members of Soldier of Fortune Magazine staff and employees are not eligible for this reward.

SOLDIER OF FORTUNE Magazine is the sole judge of any dispute arising over this reward and of any person or persons entitled to share therein and its decision on any point or matter connected with the reward shall be final and conclusive.

SOLDIER OF FORTUNE'S
AFGHAN FREEDOM FIGHTERS' FUND

Buy a Bullet, Zap a Russian Invader

All funds collected will be donated to an Afghan resistance group selected by the SOF staff.
These funds will be used to purchase arms, ammunition and medical supplies depending on the specific need of the Afghan resistance group receiving the funds.
No funds collected will be expended for salaries or administration.
Donations are NOT tax deductible.
Send your donation to:

Afghan Freedom Fighters' Fund
Box 693
Boulder, CO 80306

Copies of above poster 17"x22" available. Send $2.00 to: Afghan Poster, Soldier of Fortune Magazine, Boulder, CO 80306.

"A soldier of fortune puts his life on the line."